Quantitative Exposure Assessment

Quantitative Exposure Assessment

Stephen M. Rappaport, Ph.D.
University of California, Berkeley

Lawrence L. Kupper, Ph.D.
University of North Carolina, Chapel Hill

Published by Stephen Rappaport, El Cerrito, California, U.S.A.

Copyright@2008 by Stephen M. Rappaport

ALL RIGHTS RESERVED

No part of this publication may be reproduced, stored in a retrieval system or transmitted in any form or by any means - whether virtual, electronic, mechanical, photocopying, recording, or otherwise - without the written permission of the copyright owner.

Library of Congress Control Number: 2008900175

ISBN 978-0-9802428-0-5

Available at www.lulu.com (ID: 1341905)

To Patricia for her constant encouragement, and to the memory of Stan Roach who taught me the value of measurements.
S.M.R.

To my adorable wife Sandy, to my brother Phil, and to my friend Bob Phelan.
L.L.K.

Preface

This book describes our efforts over more than 20 years to develop a structured approach for assessing people's exposures to chemicals in occupational and environmental settings. The term *exposure* implies contact between a chemical contaminant and a portal of entry into the body, i.e. the lungs, gastrointestinal tract, or the skin. Exposure can be quantified either as the level of the contaminant at the point of contact (e.g., mg/m^3 of air in the breathing zone) or as the level of the contaminant or its product(s) inside the body (e.g., μg/l of blood or urine). Thus, the term *exposure assessment* refers to the estimation of parameters characterizing the distributions of environmental and/or biological levels of a contaminant (or its products) across an exposed population, along with attendant statistical evaluations and interpretations of such parameter estimates. Here, we regard *estimation* in the statistical sense, where each estimate of a parameter is based upon a sample of measured exposure or biological levels obtained from the population. Our approach, therefore, departs from most treatments of exposure assessment which focus either on indirect measures of exposure (questionnaire data, expert judgment, and/or deterministic models) or on the particular analytical methods used to measure chemicals in the air or other environmental media.

The topics covered in our book were the subjects of lectures presented by one of us (S. M. R.) in graduate courses taught over several years at the Schools of Public Health of the University of California, Berkeley, and the University of North Carolina, Chapel Hill, and in professional short courses throughout the world. Indeed, the writing of this book was motivated by the lack of a suitable textbook for rigorously dealing with *quantitative exposure assessment* at a level appropriate for graduate study. Students in these courses were primarily from the fields of environmental health, occupational hygiene, epidemiology, toxicology, and biostatistics.

In formulating our ideas about quantitative exposure assessment, we were fortunate to have acquired numerous longitudinal datasets from populations exposed to air contaminants. Analyses of this rich database provided important insights into the variability of chemical exposure levels received by individuals and groups over time, and motivated our emphases upon analysis of variance (ANOVA) and mixed effects models for characterizing exposure levels. Because readers of this book will have varying backgrounds and expertise with the use of such statistical models, we develop important concepts and methods chapter-by-chapter, using portions of our database for illustration. These data are available to readers, along with the particular software programs we used to perform the computations summarized in tables and figures.[1] By reproducing the results we present in the text, readers can gain confidence in their abilities to use the methods appropriately for evaluating data on exposures of interest.

We emphasize applications of exposure assessment for evaluating and controlling workplace hazards on the one hand, and for quantifying exposure-disease associations on the other. Given our interest in airborne chemical exposures, we describe procedures for collecting air samples in Chapter 1, and for setting and enforcing occupational exposure limits for air levels in Chapter 2 (including a discussion of the vagaries of exposure limits in the U.S.). Although we focus primarily upon occupational exposures, the statistical methodology we describe in Chapters 3 through 7 can be used to assess levels of chemical exposures in ambient air, water, and other environmental media of interest, and we provide references about such applications when possible. In Chapters 8 and 9, we consider the complicated milieu for evaluating and controlling occupational exposures *per se*, and we recommend particular approaches based upon our research. In Chapter 10, we show how errors of measurement in true exposure levels typically create attenuation biases in exposure-response relationships estimated from epidemiologic studies. We then demonstrate how mixed effects models, described in Chapter 6, can be used to gauge the magnitudes of such attenuation biases and to adjust estimated regression coefficients appropriately. Finally, in Chapters 11 and 12, we examine the connections between airborne exposure levels and the corresponding levels of chemicals and their products, called *biomarkers*, inside the body. We show how the relative variability of air and biomarker measurements can be used to choose an optimal strategy for quantitative exposure assessment in an epidemiologic study, using either air measurements, biomarkers, or both, as surrogates for true exposure levels.

Given the breadth of our subject matter, there is a vast literature pertaining, not only to exposure assessment *per se*, but also to the development and application of occupational exposure limits, biomarkers, statistical methods, and methods of epidemiologic analysis. We have tried to identify important early contributions and seminal papers; indeed, many of these publications are so old that they rarely show up in modern web and literature searches. Our coverage of more recent contributions is less comprehensive, due in part to the explosion during the last decade of applications of mixed effects models to exposure data. We apologize to authors of papers that we may have failed to cite.

Stephen M. Rappaport, Ph.D.
Professor of Environmental Health
School of Public Health
University of California, Berkeley

Lawrence L. Kupper, Ph.D.
Alumni Distinguished Professor of Biostatistics
School of Public Health
University of North Carolina, Chapel Hill

---

[1] Computations, which employ SAS software and Excel spreadsheets, and the datasets, can be obtained by email from S. M. Rappaport (srappaport@berkeley.edu) with proof of purchase of the book.

Acknowledgements

Many people contributed indirectly to this book. Much of our early thinking was motivated by collaborations with the late Stan Roach, who was ahead of his time in many ways, and with Steve Selvin and Bob Spear at the University of California, Berkeley. A host of Ph.D. students and post-doctoral researchers performed important analyses and provided insights along the way, notably, Hans Kromhout, Elaine Symanski, Rogelio (Mike) Tornero-Velez, Yu-Sheng Lin, Peter Egeghy, Sungkyoon Kim, Berrin Serdar, Joachim Pleil, R. C. Yu, Melissa Troester, Bob Lyles, Mark Weaver, Doug Taylor, Brent Johnson, Martha Waters, Marcie Francis, Myrto Petreas, Jenny Compton, and Ingrid Liljelind. We thank these friends for their hard work and perseverance. Finally, we thank our many colleagues who provided data for the analyses, notably Pam Susi, from the Center to Protect Workers' Rights, who supplied the welding-fume data.

Contents

1 **AIR CONTAMINANTS** 1
   1.1 Measuring airborne exposures 1
      1.1.1 Area sampling 1
      1.1.2 Breathing-zone sampling 2
      1.1.3 Personal sampling 3
   1.2 Evolution of sampling strategies 6
   1.3 This chapter and Chapter 2 8

2 **OCCUPATIONAL EXPOSURE LIMITS** 9
   2.1 Threshold Limit Values 9
      2.1.1 Long-term versus short-term TLVs 9
      2.1.2 Health basis of TLVs 11
   2.2 OSHA standards 14
      2.2.1 Risk and feasibility 14
      2.2.2 Interpretation of PELs 16
   2.3 Working limits 17
   2.4 This chapter and Chapter 3 18

3 **SAMPLING EXPOSURES** 19
   3.1 Establishing observational groups 19
      3.1.1 Example 20
   3.2 Samples of data 21
   3.3 Random versus worst-case sampling 22
      3.3.1 Random sampling 23
   3.4 Self assessment of exposure 23
   3.5 This chapter and Chapter 4 25

4 **EXPOSURE DISTRIBUTIONS** 27
   4.1 Exposure as a random process 27
   4.2 Lognormal distributions of exposures 29
   4.3 Stationarity 32
   4.4 Autocorrelated exposure series 33
   4.5 Measurements below the limit of detection 35
   4.6 Estimation of parameters 35
   4.7 Normality of (logged) individual exposure levels 36
   4.8 This chapter and Chapter 5 37

5 **EXPOSURE VARIABILITY WITHIN AND BETWEEN PERSONS** . 39
   5.1 Importance of between-person variability 39
   5.2 One way random effects model 40
      5.2.1 Assumptions 41
      5.2.2 Lognormal distributions 43
      5.2.3 Estimating parameters 45
      5.2.4 The ANOVA table 47
   5.3 Relative measures of variability 47
   5.4 Ranges of variance components in occupational groups 49

- 5.4.1 Sources of variability in occupational groups 50
- 5.4.2 Uniformity of mean exposure levels across persons 52
- 5.5 Environmental exposures 52
- 5.6 This chapter and Chapter 6 54

6 MIXED MODELS OF EXPOSURE 55
- 6.1 General linear mixed models 55
  - 6.1.1 Matrix notation 55
- 6.2 Fitting linear mixed models separately to Groups 1 – 4 57
  - 6.2.1 Profile plots 57
  - 6.2.2 REML estimates 58
  - 6.2.3 Normality of predicted random effects 59
  - 6.2.4 Normality of residuals 60
- 6.3 Modeling data involving several groups and covariates 62
  - 6.3.1 Assumptions 63
  - 6.3.2 Estimating parameters 64
  - 6.3.3 Estimating variance components across groups 64
  - 6.3.4 Combining data from multiple groups 66
- 6.4 This chapter and Chapter 7 67

7 DETERMINANTS OF EXPOSURE LEVELS 69
- 7.1 Preliminary investigation 69
- 7.2 Profile plots 72
- 7.3 Application of mixed models 72
  - 7.3.1 Normality of standardized random effects and residuals 73
- 7.4 Investigating covariates 75
  - 7.4.1 Exposures predicted from fixed effects 76
  - 7.4.2 Modeling the random effects 78
- 7.5 Conclusions 79
- 7.6 This chapter and Chapter 8 80

8 PROBABILITIES OF EXCEEDING OELs 81
- 8.1 Pitfalls of compliance testing 81
- 8.2 Risk and overexposure 83
- 8.3 Relating probabilities to exposure distributions 84
  - 8.3.1 Estimating exceedance and probability of overexposure 86
  - 8.3.2 A large sample of observational groups 86
- 8.4 Compliance vs. health risk 87
- 8.5 Using exceedance to characterize acceptable exposure 89
- 8.6 Using the group mean to characterize acceptable exposure 91
  - 8.6.1 Connections to probabilities of exceeding the OEL 92
  - 8.6.2 Application to STELs 94
- 8.7 Regulatory implications 94
- 8.8 This chapter and Chapter 9 95

9 INTEGRATING EXPOSURE ASSESSMENT WITH CONTROL 97
- 9.1 An integrated strategy 97
- 9.2 Assessing acceptable exposure levels 99
  - 9.2.1 Testing exposure for a group 100

    9.2.2 Sample-size requirements 101
    9.2.3 Alternative test when the estimated between-person variance
    component is zero 103
    9.2.4 Testing for overexposure 104
  9.3 Selecting appropriate controls 106
  9.4 Conclusions 108
  9.5 This chapter and Chapter 10 109

10 EXPOSURE MEASUREMENT ERRORS 111
  10.1 Exposure-response relationships 111
  10.2 Individual-based studies 113
    10.2.1 Regression analysis 113
    10.2.2 Estimating sample sizes 114
  10.3 Group-based studies 115
    10.3.1 Adding a random group effect 115
    10.3.2 Health-outcome model 117
    10.3.3 Regression analysis 118
    10.3.4 Estimating sample sizes 119
  10.4 Adjusting estimated regression coefficients 120
  10.5 Comparing individual-based and group-based studies 121
  10.6 Dichotomous health outcomes 122
    10.6.1 Regression analysis 123
  10.7 Summary 124
  10.8 This chapter and Chapter 11 125

11 EXPOSURE, DOSE, AND DAMAGE 127
  11.1 Processes relating exposure and disease 127
  11.2 Linear kinetics 128
  11.3 The concept of dose 129
  11.4 Burden and dose for 'on-off' exposures 130
    11.4.1 'On-off' exposure to styrene 132
    11.4.2 Time to steady state 133
  11.5 Random exposure 133
    11.5.1 Dose following random exposure 135
    11.5.2 Rapid versus slow elimination 135
  11.6 Physiological damping of exposure variability 137
  11.7 Occupational exposure to mercury 139
  11.8 Occupational exposure to styrene 140
  11.9 Extending the concept of dose 141
  11.10 This chapter and Chapter 12 142

12 BIOMARKERS OF EXPOSURE 143
  12.1 Definitions of biomarkers 143
  12.2 Time scales of biomarkers 144
  12.3 Choosing between air measurements and biomarkers 145
    12.3.1 Intrinsic advantages 145
    12.3.2 Biasing effects of surrogate exposure measures 147
    12.3.3 Estimated variance components 147

12.3.4 Estimated variance ratios 148
12.3.5 Estimated lambda ratios 150
12.4 Choosing between air measurements and biomarkers 153
12.5 This chapter 154
Bibliography **157**
Index **165**

1 AIR CONTAMINANTS

1.1 Measuring airborne exposures

The endeavors we now associate with the field of *quantitative exposure assessment* developed largely from the awareness that workers exposed to increasing levels of some air contaminants were at increased risks of developing occupational diseases. Although it was known in ancient times that workers exposed to certain dusts and heavy metals suffered health effects as a consequence, studies linking exposure levels with health risks were not reported until the early part of the 20th century. Certainly by the 1920s, levels of air contaminants in mines and factories had been documented. This work ultimately gave rise to the field of *occupational hygiene* (also called *industrial hygiene*), governmental entities devoted to factory inspections, and occupational exposure limits (OELs). In the 1950s, much of the interest in quantitative exposure assessment extended beyond the workplace to cities where noxious contaminants in the air (and their effects) were increasingly apparent; this motivated parallel developments of training programs for air hygiene, governmental inspectorates, exposure limits, etc.

Because most investigations of exposures to chemical contaminants have focused upon the airborne route, it is useful to preface our discussion of quantitative exposure assessment with a brief overview of historical changes in the techniques for measuring air contaminants *per se*. For a concise synopsis of the history of air sampling techniques, see the commentary by Cherrie (2003). Instrumentation and methods for air sampling are discussed in detail by Cohen and McCammon (2001).

1.1.1 Area sampling

During the early days, equipment for measuring air levels was cumbersome. Because the apparatus was bulky and generally required electric pumps to draw air through a collector, air monitoring was conducted at fixed sites in the workplace; thus, the measurements were often referred to as *area* or *static* samples. After trapping gaseous or particulate contaminants from a known volume of air, samples were taken to a laboratory where the contaminant mass (or number of particles) was determined. Occupational hygienists used area sampling along with information about workers' locations to estimate exposure levels. Area sampling is still employed for most measurements of ambient air pollutants.

Figure 1.1 illustrates some uses of area sampling for measuring occupational exposures. Figures 1.1A and 1.1C show early air samplers for benzene vapors. The first method employed activated carbon to adsorb benzene from the air (Figure 1.1A) for subsequent gravimetric measurement (Greenburg, 1926), while the second (Figure 1.1C) used concentrated nitric/sulfuric acids to convert benzene to nitrobenzene for subsequent titration with stannous chloride (Smyth and Smyth, 1928). Figure 1.1B shows a Greenburg-Smith impinger, which was used to collect dusts as described in the 1936 monograph *Industrial Dust* by Drinker and Hatch (1936). Air was drawn at 28 l/min (one ft^3/min) through a glass vessel (about 1-liter volume) where particles were trapped by impingement against a glass plate and subsequently suspended in water. In the laboratory, a drop of the water containing the suspended particles was examined under a light microscope to count the particles. In Figure 1.1D, a high-volume sampler and a cascade impactor were used to collect airborne asbestos in a packing operation (about 1953). The dust was subsequently measured in a laboratory, either by weighing or by counting/sizing particles with a microscope.

1.1.2 Breathing-zone sampling

Sometime after the advent of area sampling, say in the late 1940s [although earlier applications were mentioned, e.g., Smyth and Smyth (1928)], *breathing-zone* sampling was performed, where air levels were measured in the immediate vicinity of the worker for short periods of time. The inspector would position a flexible probe or small collector at head height, as close as practical to the worker, while tasks were performed. Two illustrations of breathing-zone sampling are shown in Figure 1.2. At left, a midget impinger was used to collect explosive vapors (probably nitroglycerine) in a U.S. factory producing ordinance during World War II. At right, benzene vapors were collected from the breathing zone of a worker producing mechanical seals with both a direct reading instrument (explosimeter) and tubes containing silica gel. Generally, 3 or 4 breathing-zone samples were obtained during a work shift to determine contaminant levels at different times or for particular tasks (Breslin et al., 1967). Thus, air concentrations were weighted by the corresponding exposure times to estimate the average exposure over the full day; this was referred to as the 8-*hour (h) time-weighted-average (TWA) air concentration*. For example, if benzene levels were measured at 0.40 ppm for two h, followed by 0.60 ppm for 4 h, and 0.10 ppm for 2 h, then the 8-h TWA is computed as:

$$\text{8-h TWA} = \frac{(0.40 \text{ ppm} \times 2 \text{ h}) + (0.60 \text{ ppm} \times 4 \text{ h}) + (0.10 \text{ ppm} \times 2 \text{ h})}{(2 \text{ h} + 4 \text{ h} + 2 \text{ h})} = 0.42 \text{ ppm}$$

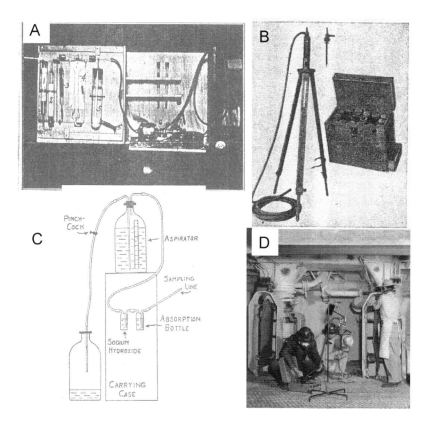

Fig. 1.1 Area sampling for air contaminants. A) Collection of benzene by adsorption on charcoal about 1920 (Greenburg, 1926). B) Greenburg-Smith impinger used for dust sampling about 1930 (Drinker and Hatch, 1936). C) Collection of benzene by aspiration and absorption in acid about 1925 (Smyth and Smyth, 1928). D) Collection of asbestos with a high-volume sampler about 1950. (Photograph courtesy of the National Institute for Occupational Safety and Health).

1.1.3 Personal sampling

With the development of small battery-operated pumps around 1960, it became possible to affix the air sampler to the worker's clothing in the breathing zone and thus to collect *personal* samples over the entire work shift (Sherwood and Greenhalgh, 1960). Figure 1.3 shows early prototypes of these pumps, developed by R. J. Sherwood in the late 1950s (Sherwood, 1997). Symanski *et al.* (1998b) reported that one-half of surveys of chemical exposures reported between 1967 and 1996 employed either personal sampling (44%) or a combination of personal and area sampling (6%). Comparisons of results from personal and area sampling in the workplace indicate that personal sampling finds consistently greater air levels than area sampling and that the ratio of personal/area measurements increases with room size and the quality of ventilation [reviewed by Cherrie (2003)].

 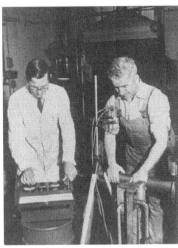

Fig. 1.2 Two examples of breathing-zone sampling. Left: Sampling with a midget impinger of vapors (probably nitroglycerine) at an ordinance plant – U.S. (1943). (Photograph courtesy of the National Institute for Occupational Safety and Health). Right: Benzene sampling with an explosimeter and silica gel tubes during manufacture of mechanical seals – U.K. (1950). (Photograph courtesy of R.J. Sherwood).

Today, virtually all measurements of occupational exposure involve personal sampling over the full work shift. Samplers can utilize either *active monitors* (i.e., those that use small pumps) or *passive monitors* that rely upon diffusion to transfer the contaminant to the collector (housed in a badge or tube). Examples of personal sampling with active and passive monitors are shown in Figure 1.4. Because passive monitors are simple and convenient to use, they make it possible to collect many more measurements than can be obtained with active monitors.

Despite the widespread availability of personal monitors since the early 1960s, and the extensive use of personal sampling of occupational exposures, it is surprising that so few studies of ambient exposures to air pollutants have employed personal sampling. Indeed, personal sampling of ambient air pollutants was essentially an unused technology prior to 1990, aside from the classic Total Exposure Assessment Methodology (TEAM) studies conducted by the U.S. EPA in the early 1980s [reviewed by Rappaport and Kupper (2004)]. When personal sampling has been conducted for studies of ambient air pollutants, the variability of ambient exposure levels has been shown to increase several fold over that indicated by area sampling (usually referred to as *microenvironmental sampling* in this context) [e.g., see Spengler *et al.* (1994)], consistent with results from occupational studies.

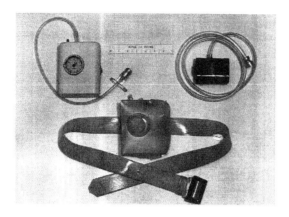

Fig. 1.3 Prototypes of the first personal samplers, used between 1955 and 1958. Battery-operated pumps were housed in convenient receptacles, i.e., small jewelry cases (upper left and bottom) and a bicycle-lamp case (upper right). [From *Applied Occupational and Environmental Hygiene*, Realization, Development, and First Applications of the Personal Air Sampler, 12(4): pages 229–234. Copyright 1997. ACGIH®, Cincinnati, OH. Reprinted with permission.]

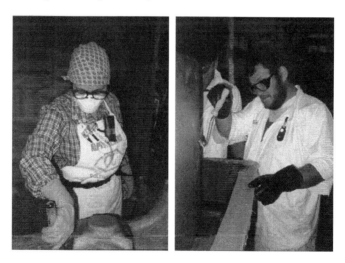

Fig. 1.4 Personal samplers used to measure styrene and other contaminants in the reinforced-plastics industry (1986). Left: Active sampling with sorbent tubes and a micro-impinger (the battery-operated pump is worn at the waist (not visible). Right: Passive sampling with a badge containing activated carbon.

1.2 Evolution of sampling strategies

Aside from the important technological advances that permitted air measurements to more accurately reflect personal exposures, the underlying strategy for air monitoring also changed radically from the 1920s to the present time. In the period between 1920 and 1950, only a handful of health professionals were assessing chemical exposures, and these practitioners were largely employed by governmental agencies, academic institutions, and insurance companies. As noted previously, the air monitoring equipment during that period was cumbersome, averaging times were only a few minutes in duration, and assays were nonspecific and lacked sensitivity. These limitations resulted in relatively few quantitative studies of exposure levels and their determinants. Yet, despite the primitive state of the measurement art, investigators routinely collected large numbers of measurements in their studies, sometimes several hundred (Ashford, 1958; Bloomfield and Greenburg, 1933; Oldham and Roach, 1952; Smyth and Smyth, 1928). A likely motivation for such large samples sizes was the recognition that air levels varied tremendously over time. Since the primary focus of early assessments of exposure was to quantify the relationship between chemical exposure and the risk of disease, large numbers of air measurements were needed to accurately estimate the average concentrations experienced by different groups of workers.

In a classic early study, Oldham and Roach explored the link between coal workers' pneumoconiosis and exposure to coal dust in British mines around 1950 (Oldham, 1953; Oldham and Roach, 1952; Roach, 1953). Breathing-zone measurements were obtained with a thermal precipitator, a heavy device, which was operated for only 3 minutes at a time. Dust levels were repeatedly monitored for each coal worker during a two-day survey at up to 20 randomly selected times per worker (779 measurements in all). The variability of exposure at different times is evident in Figure 1.5, which reproduces a portion of the appended data from that investigation (Oldham and Roach, 1952). For example, collier (coal worker) No. 6 experienced coal dust levels ranging from less than 100 to 2,000 particles per milliliter, more than a 20-fold range on a single day!

With routine application of personal monitors after 1970, occupational hygienists could measure individual exposures relatively easily over the full work shift. In so doing, they quickly learned that the dramatic variability of exposure measurements, which had been observed for short periods of breathing-zone sampling, extended to shift-long levels as well. Also, after passage of the Occupational Safety and Health Act (OSH Act) of 1970 (OSH-Act, 1970), the numbers of occupational hygienists increased rapidly in the U.S. Thus, it became possible to obtain large numbers of full-shift personal measurements for the many populations exposed to hazardous air contaminants.

APPENDIX
RESULTS OF "RANDOM COLLIERS" SURVEY

Collier	Total Time Spent on Coalface (min.)	Duration of Mid-shift Break (min.)	Thermal Precipitator Samples (No. of particles per ml. between 0·5 and 5·0μ)										
1	315 430	25 15	1,630 950	800 <100*	1,100	920	1,500	1,770	980	1,430	1,540		
2	—	—	Collier no longer working at pit										
3	350 365	20 20	<100 800	1,320 1,180	650 1,130	<100 880	<100 430	360 430	(<100)† 390	<100 360	220 460	180 440	
4	345	20	1,750	1,690	2,030	2,730	1,980	940	1,360	2,070	2,190	830	2,650
5	— —	— —	} Collier absent on both occasions										
6	270 335	20 25	570 240	<100 170	<100 450	(210) 560	310 320	740 280	2,000 (<100)	520 320	520	820	520 510
7	325 365	40 35	390 1,110	700 730	920 (<100)	1,240 (<100)	1,580 750	1,530 1,090	920 720	530 1,130	580	770	
8	300 395	25 25	580 350	770 1,460	470 800	(<100) 1,250	1,150	1,010	(<100)	860	920	760	840

Fig. 1.5 A portion of an appendix from a paper by Oldham and Roach (1952). Each entry represents the dust level for a random 3-min sample obtained from the breathing zone of a collier (coal worker). Note that several measurements were obtained from each subject on a given day. (*Br J Ind Med* 1952; 9: 112-119; reproduced with permission from the BMJ Publishing Group).

Yet, this movement towards collection of large numbers of personal measurements was not realized. In a review of 564 papers published on the etiology of chronic diseases in the *American Journal of Epidemiology* during the 1980's, only 13% of the studies referred to any quantitative measurements of exposure levels (Armstrong *et al.*, 1992). Furthermore, in studies that did measure exposure levels, sample sizes were very small. Based upon 696 data sets, collected primarily from the epidemiologic literature between 1967 and 1996, Symanski *et al.* (1998b) reported that the median sample size was 4 measurements and that only 21% of studies had more than 10 measurements. Although information is rare regarding the numbers of exposure measurements collected during private industrial inspections, Tornero-Velez *et al.* (1997) reported sample sizes for 4,864 annual surveys obtained in the nickel producing industry during the period of 1970 – 1990. Here, the median number of measurements was one, 85% of surveys had 4 measurements or fewer, and 95% had 10 or fewer, consistent with the results of Symanski *et al.* (1998b). Clearly, such small sample sizes are inadequate to accurately estimate the important parameters of exposure distributions in the face of great variability. And, sadly, the situation may be getting worse rather than better, given indications that even fewer measurements have been collected after 1990 (Cherrie, 2003).

So, despite the relentless advances in measurement technology and the influx of large numbers of occupational hygienists, the numbers of exposure measurements per survey did not increase and may have actually diminished in the latter part of the 20th century. This trend seems to reflect a fundamental change in focus for occupational exposure assessment. Rather than obtaining data with which to link health effects and exposures, occupational hygienists appear to have increasingly devoted their efforts to evaluating *compliance* with occupational exposure limits (OELs).

Whereas exposure limits were essentially nonexistent prior to 1950, OELs for more than 400 chemical substances had been developed by 1971 when the OSH Act established these levels as legally enforceable limits in the U.S. Indeed, evidence of noncompliance carried with it sanctions against the employer under the OSH Act. Thus, the *de facto* role of the occupational hygienist, the vast majority of whom were privately employed, became one of limiting corporate liability. By shifting focus to the possibility that exposure levels could exceed the OEL during a single work shift, the sampling time frame became very short (days), there were few measurements collected, and coverage was limited to workers thought to have the greatest exposures (*worst-case sampling*).

1.3 This chapter and Chapter 2

In this chapter we noted that the field of quantitative exposure assessment began with studies of workers exposed to toxic air contaminants. Thus, the methods for quantifying exposures have been tied largely to technological and regulatory developments in the area of occupational health. Interestingly, with passage of the OSH Act in 1970, it appears that the motivation for assessing chemical exposures shifted from studies designed to elucidate the health impacts of exposures to surveys of compliance with OELs. If compliance with OELs did, indeed, alter the motivation for exposure assessment, it is important to understand the mechanisms by which OELs have been set and enforced in the U.S. This will be the subject of Chapter 2.

2 OCCUPATIONAL EXPOSURE LIMITS

Since most assessments of workplace exposures have arguably been motivated by the need to comply with OELs, some understanding of the basis for these limits is important to the proper assessment of exposure. The following analysis focuses upon exposure limits used in the U.S. OELs in other countries share many features with the U.S. limits. For a more comprehensive review of the history and basis for OELs, see Paustenbach (2000).

2.1 Threshold Limit Values

Prior to passage of the Occupational Safety and Health Act (OSH Act) of 1970 (OSH-Act, 1970), limits for levels of airborne contaminants in the U.S. generally reflected consensus standards developed by various groups, most notably the Threshold Limit Values (TLVs) of the American Conference of Governmental Industrial Hygienists (ACGIH) (ACGIH, 2007). The first list of TLVs was published in 1946 for 150 chemical substances (LaNier, 1984) and has been updated annually since then to include more than 600 substances today. The TLVs have had a profound influence on the practice of occupational hygiene throughout the world (Rappaport, 1993b). The U.S. government (through the Occupational Safety and Health Administration, OSHA), several states within the U.S., and many other countries have adopted some or all of the TLVs as their official limits.

2.1.1 Long-term versus short-term TLVs

The ACGIH defines TLVs according to criteria for monitoring over both the long-term and the short-term. The long-term limit, referred to as the TLV-TWA (time-weighted average), is defined as (ACGIH, 2007):

> "... the TWA concentration for a conventional 8-hour workday and 40-hour workweek, to which it is believed that nearly all workers may be repeatedly exposed, day after day, for a working lifetime without adverse effect.".

By defining a TLV-TWA as the average air concentration to which a worker can be repeatedly exposed over a working lifetime, the ACGIH suggests that this is the average workplace concentration to which a worker could be exposed regardless of fluctuations in air levels over time.² However, such an interpretation is equivocal because the ACGIH also makes recommendations regarding excursions of the TLV-TWA during the work shift (Hewett, 1997a).³

The ACGIH provides a second set of OELs that limit air concentrations over short periods of time. These are referred to as *short-term exposure limits* (STELs), or TLV-STELs, representing air concentrations averaged over 15 min, to which

> "...workers can be exposed continuously for a short period of time without suffering from 1) irritation, 2) chronic or irreversible tissue damage, 3) dose-rate-dependent toxic effect, or 4) narcosis... , and provided that the daily TLV-TWA is not exceeded. The TWA-STEL is not a separate, independent exposure guideline, rather it supplements the TWA limit where there are recognized acute effects from a substance whose toxic effects are primarily of a chronic nature" (ACGIH, 2007).

Two of these four criteria, that is, irritation and narcosis, are consistent with the accepted perception that STELs relate to effects of transient high levels that might arise even though the average concentration is at or below the TLV-TWA (Henschler, 1984; Ulfvarson, 1987; Zielhuis *et al.*, 1988). Likewise, the fact that irreversible tissue damage, such as neural death associated with anoxia, might be associated with periods of intense exposure provides a clear rationale for STELs. However, the notion that long-term damage might depend on the magnitude of short-term fluctuations in air levels at a given TLV-TWA (i.e., "dose-rate-dependent toxic effects") is more controversial. This suggests that transmission of the contaminant from the air to the target tissues is rapid and that the relationship between tissue damage and exposure is nonlinear (concave upward) (Rappaport, 1985; Rappaport, 1991b). Little, if any, evidence of such behavior has been reported for situations in which the average exposure is less than the TLV-TWA.

A third type of limit, designated the TLV-C, represents a *ceiling* value that should not be exceeded even instantaneously. A review of the TLVs (ACGIH, 2007) suggests that ceiling limits are applied to substances that only produce acute effects.

---

² Logically, a 'conventional' workday or workweek would reflect average conditions and not those associated with unusually high or low exposures. Likewise, the wording 'repeatedly exposed, day after day' suggests a cumulative exposure (i.e., average exposure times duration of exposure). However, the ACGIH has not provided guidance as to the exact meaning of the TLV-TWA when confronted with variability of air levels experienced by a given worker from day to day.

³ The ACGIH suggests that "... worker exposure levels may exceed 3 times the TLV-TWA for no more than a total of 30 minutes during a workday, and under no circumstances should they exceed 5 times the TLV-TWA, provided that the TLV-TWA is not exceeded."

2.1.2 Health basis of TLVs

The ACGIH's definition of its TLV-TWAs as levels that protect "...nearly all workers repeatedly exposed, day after day ..." implies that these limits are based primarily on health considerations. For many agents, the paucity of exposure-health response data available to, or reported by, the ACGIH has caused this interpretation to be questioned (Halton, 1988; Henschler, 1984; Roach and Rappaport, 1990; Zielhuis *et al.*, 1988). The documentation supporting a particular TLV provides a relatively brief literature review (ACGIH, 2001). Although it contains no formal analysis of exposure-response relationships, the documentation may comment upon such analyses from other sources.

Some TLVs are based on animal experiments, others on reports of human experience both in the workplace and in controlled experiments with volunteers. Referring to the TLVs based upon human experience, Roach and Rappaport reported that the percentages of humans suffering adverse health risks were surprisingly large at the exposure concentrations represented by the 1976 and 1986 TLVs, averaging between 14 and 17 subjects per 100 exposed individuals (Roach and Rappaport, 1990). This is seen in the left portion of Figure 2.1, which shows the 1976 TLV-adjusted average exposure level reported in a study cited by the ACGIH versus the percent of adversely affected persons in that study (effects ranged from short-term irritation to long-term morbidity and death). The horizontal dashed line represents the case where the average exposure equaled the TLV and the solid line represents the best least-squares fit to the data (in log scale). Since only a small amount of the variability in the (logged) TLV-scaled exposure level was explained by the percentage of persons adversely affected ($R^2 = 0.063$), Roach and Rappaport concluded that considerations of health could not have been the primary motivation for the TLVs. They further observed that the TLVs were highly correlated with the average levels of exposure reported in these same studies cited by the ACGIH. As shown in the right portion of Figure 2.1, the 1976 TLVs explained about 70% of the variability in average exposure levels reported in the cited studies ($R^2 = 0.700$). Roach and Rappaport speculated, therefore, that the TLVs historically reflected the perception of the ACGIH's TLV Committee that more stringent limits were unrealistic given the perceived state of control at the time. In fact, the right portion of Figure 2.1 shows that the least-squares fit to the data (solid line) lies above the 45-degree line representing strict equality (dashed line) and suggests that TLVs were historically somewhat lower on average than the exposure levels reported in the studies. Furthermore, the difference between the average exposure level and the corresponding TLV increased as the TLV became smaller. Because substances with smaller TLVs tend to be more toxic, this suggests that the difference between the TLV and the average exposure level was greater for the more toxic substances. For example, a typical 1976 TLV of 0.1 (arbitrary units) was about 1/5th of the average exposure level, while a typical 1976 TLV with a value of 100 was about half of the average exposure level.

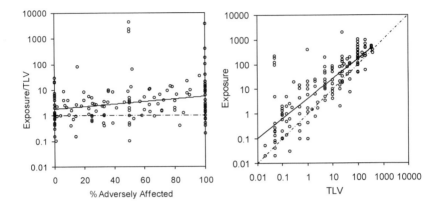

Fig. 2.1 Relationships derived from human studies cited by the ACGIH in its documentation for the 1976 TLVs. Left: The average exposure expressed as a multiple of the TLV (Y) versus the percentage of persons who experienced adverse effects (X). The horizontal dashed line represents the case where average exposure equaled the TLV and the solid line represents the best least-squares fit to the data ($\hat{Y} = 1.71e^{0.011X}$; $R^2 = 0.063$). Right: The average level of exposure (Y) versus the TLV (X). The 45-degree dashed line represents the case where average exposure equaled the TLV and the solid line represents the best least-squares fit to the data ($\hat{Y} = 3.93X^{0.806}$; $R^2 = 0.700$). [Data from Roach and Rappaport (1990)].

The conclusion that TLVs were strongly influenced by actual workplace exposures draws support from statements of prominent industrial toxicologists in the 1950s and 1960s and by the staged reductions in TLVs, which have occurred in step with general trends toward lower exposures (Rappaport, 1993b). Such reductions are illustrated in Table 2.1 for ACGIH TLVs set for benzene over the years. This conclusion is also consistent with the notion, expressed by Castleman and Ziem (1988), that corporate interests had exerted undue influence on the setting of TLVs.

The critiques of the ACGIH TLVs by Castleman and Ziem (1988) and Roach and Rappaport (1990) were controversial, to say the least, and many of the conclusions were vigorously debated [see Paustenbach (2000)]. Nonetheless, after publication of these papers in 1988 and 1990, the ACGIH instituted changes regarding the composition and workings of its TLV Committee. After investigating the list of intended changes in the 1991-1992 list of TLVs, Rappaport speculated that the TLV Committee was applying more stringent health criteria for setting limits after 1989, particularly for carcinogenic substances (Rappaport, 1993b). That observation was based upon fold reductions for TLVs of recognized carcinogens that were much greater after 1989 (median reduction = 7.5-fold) than previously (median reduction = 2.0-fold to 2.5-fold), as suggested in Table 2.1 for benzene. Smith and Mendeloff confirmed this finding in 1999 by reporting that the estimated regression coefficient relating the magnitude of a TLV to the risk of cancer was significantly greater after 1989 than before 1989 (Smith and Mendeloff, 1999).

Table 2.1 Historical reductions in ACGIH TLVs for benzene.

Year	TLV (ppm)	Fold Reduction
1946	100	
1947	50	2.0
1948	35	1.4
1957	25	1.4
1974	10	2.5
1997	0.5	20.0

Because TLVs have been updated rather infrequently, the notion that many of them represent conditions in the distant past should be troubling to persons interested in occupational health. As shown in Figure 2.2, the ages of the 1991-1992 TLVs ranged from 0 to 45 y with a median age of 16.8 y. Given the consistent trends towards lower exposures noted earlier, documented by Symanski and coworkers (Symanski *et al.*, 1998a; Symanski *et al.*, 1998b), this suggests that many TLVs exist at levels much greater than current exposures. Interestingly, Symanski *et al.* presented evidence that the rate of reduction in exposure from year to year was less for substances having one or more changes in the TLV, perhaps because the newer TLVs were thought to represent 'safe' levels.

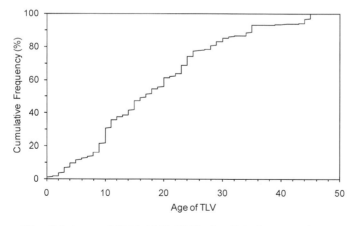

Fig. 2.2 Ages of 1991-1992 TLVs for 625 air contaminants. The median age was 16.8 y [data from Rappaport (1993b)].

2.2 OSHA standards

The OSH Act established the Occupational Safety and Health Administration (OSHA) as the official standard-setting body in the U.S. (OSH-Act, 1970). The *Permissible Exposure Limits* (PELs) established by OSHA for chemical agents fall into two categories, namely, those originally adopted as existing standards in 1971 under a provision of the OSH Act, and those subsequently developed as parts of new standards. Those adopted as existing standards included PELs for about 400 chemicals, most of which had originally been issued as TLVs by the ACGIH in 1968. As shown in Table 2.2, OSHA has developed new exposure limits for only 16 substances. The new standards often included a second exposure criterion called an *Action Level* (AL), an exposure concentration of roughly one-half of the PEL (which triggers certain actions), and sometimes a STEL (typically 5-times the level of the PEL over 15 min).

Table 2.2. New PELs promulgated by OSHA.

Substance	PEL	AL	STEL
Acrylonitrile	2 ppm	1 ppm	10 ppm
Arsenic	10 µg/m^3	5 µg/m^3	None
Asbestos	0.1 fiber/cm^3	None	1.0 fiber/cm^3
Benzene	1 ppm	0.5 ppm	5 ppm
1,3-Butadiene	1 ppm	0.5 ppm	5 ppm
Cadmium	5 µg/m^3	2.5 µg/m^3	None
Coke oven emissions	150 µg/m^3	None	None
Cotton dust	200 µg/m^3	None	None
Chromium (VI)	5 µg/m^3	2.5 µg/m^3	None
1,2-Dibromo-3-chloropropane	1 ppb	None	None
Ethylene oxide	1 ppm	0.5 ppm	5 ppm
Formaldehyde	0.75 ppm	0.5 ppm	2 ppm
Lead	50 µg/m^3	30 µg/m^3	None
Methylene chloride	25 ppm	12.5 ppm	125 ppm
Methylenedianiline	10 ppb	5 ppb	100 ppb
Vinyl chloride	1 ppm	0.5 ppm	5 ppm

2.2.1 Risk and feasibility

Court decisions over the period between about 1975 and 1990 have determined that OSHA may issue a new standard after demonstrating that workers are at *significant risk* of adverse health effects at the level of the existing PEL. Significant risk has been interpreted by OSHA, in situations involving carcinogens, to mean that no more than 1 worker out of 1000 exposed at the level of the PEL over a 45-year working lifetime would develop a specific type of cancer (OSHA, 1987). Having demonstrated a significant risk at an existing limit, OSHA establishes a new PEL representing the lowest level thought to be economically and technologically feasible in at least one industrial sector (Rappaport, 1993b). If the residual risk at this new PEL still exceeds 1 per 1000, then the door is left open for OSHA to institute further PEL reductions in the future. This explicit consideration of risk and feasibility differentiates OSHA's new PELs from TLVs and apparently from many other OELs promulgated elsewhere (Henschler, 1984; Zielhuis et al., 1988).

The two-tiered process of setting new PELs is illustrated in Figure 2.3 using the benzene standard (OSHA, 1987; Rappaport, 1991b) as an example; this process has been used for most of the 16 new PELs that are carcinogens. For its benzene standard, OSHA devoted a large section of the preamble to assess the risks at the old and new PELs. Using the results from three epidemiologic studies to arrive at an exposure-response relationship, OSHA applied a linear cancer model to quantify the risks associated with the old and new PELs. Assuming a cumulative exposure to benzene at the old PEL (10 ppm), under a regimen of 8 hr/day, 5 day/week for 45 years, OSHA estimated the risk to be 95 excess deaths from leukemia per 1000 exposed workers. Clearly, such a large risk would be well above the one per 1000 level that was needed to justify a new PEL. Then OSHA set a new PEL of 1 ppm, which was argued to be the lowest level that was feasible to achieve (discussed in the next paragraph). Note from Figure 2.3 that even the residual risk inherent in the *new* PEL, i.e., 9.5 deaths per 1000 workers, was well above the one per 1000 trigger needed to justify a new standard. Such a large residual cancer risk at the level of an OEL is not unusual (Alavanja et al., 1990).

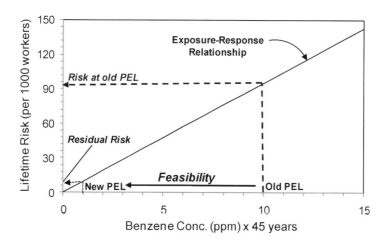

Fig. 2.3 Process used by OSHA to establish the risk of leukemia at the old and new PELs for benzene. Note that the reduction in risk is driven by feasibility because even the new PEL poses a significant risk of leukemia (i.e., greater than one per thousand).

Having demonstrated the need for a new standard, based upon health-risk, OSHA then endeavored to determine the lowest level of benzene exposure that was *feasible* based upon on both technologic and economic factors. The preamble to OSHA's benzene standard contained a lengthy discussion of the levels of benzene exposure in various industrial sectors, i.e., petrochemical production, petroleum refining, coke- and coal-chemical production, tire manufacturing, bulk terminals and plants, and transportation (via tank trucks). OSHA subsequently determined that, in fact, only petrochemicals and coke- and coal-chemical production could not feasibly achieve exposures below one ppm (OSHA, 1987; Rappaport, 1993b). Thus, it seems that OSHA developed its new benzene PEL of one ppm relative to the *two worst sectors of industry* affected by the standard at the time of its promulgation. Surprisingly, OSHA's data indicated that these same two sectors contained only 2.2% of all workers exposed to benzene and only 9.1% of all workers apparently exposed above one ppm (Rappaport, 1993b) (Table 2.3). Thus, the two smallest industrial sectors dictated the official limit for the vast majority of workers who were, in fact, employed elsewhere, including 91% of the workers who were apparently exposed above one ppm.

Table 2.3. Estimated numbers of workers exposed to benzene in the U.S. at air levels relative to the OSHA PEL of 1 ppm [from Rappaport (1993b)].

Industrial Sector	Exposure (ppm)	No. Workers Exposed	No. Workers Exposed >1 ppm
Petrochemicals	0.728	4,300	619
Petroleum refining	0.228	47,547	2,045
Tire manufacturing	0.204	65,000	2,470
Bulk terminals and plants	0.223	72,418	3,042
Transportation (tank truck)	0.199	47,600	1,714
Coke and coal chemicals	1.367	947	313
TOTAL		237,812	10,202

2.2.2 Interpretation of PELs

Since OSHA equates risk with cumulative exposure over 45 years (i.e., the average exposure concentration times 45 years) and considers feasibility in the context of average air levels, each new PEL arguably represents the average exposure concentration received by the individual worker over 45 years in those segments of industry considered least able to control exposures. This interpretation is supported by analyses (noted above) indicating that, at the time the benzene standard was set, the petrochemicals and coke- and coal-chemical sectors had overall mean exposure concentrations of 0.728 and 1.367 ppm, respectively, levels close to the new PEL of one ppm (see Table 2.3) (Rappaport, 1993b). Thus, the philosophy used to establish a new PEL is generally consistent with the idea that a person's risk is proportional to his or her average exposure concentration over a fixed period of time (e.g., 45 years). OSHA also noted that some segments of industry affected by the benzene standard could consistently maintain exposure concentrations below the new PEL and that workers would experience proportionally lower risks (OSHA, 1987); here again, individual risk was assumed to be proportional to the person's average exposure concentration over 45 years.

Notwithstanding the fact that the standard-setting process used by OSHA equated risk with cumulative exposure (i.e., average exposure level times duration of exposure) over 45 years, the definition used by the agency for enforcement is ambiguous. For example, the benzene standard defined the PEL in terms of the employer's responsibility as follows (OSHA, 1987)[4]:

"The employer shall assure that no employee is exposed to an airborne concentration of benzene in excess of 1 ppm as an 8-hour time-weighted average."

While this statement can be taken to mean that no employee should be exposed to more than one ppm of benzene *on the average*, such an interpretation was initially rejected (OSHA, 1987). However, in a subsequent ruling regarding the feasibility of its standard for inorganic lead, OSHA indicated that it will not issue a citation on the basis of a single measurement above the PEL, but rather requires that exposures be less than the PEL "... *in most operations most of the time...*" (italics added) (OSHA, 1991)[5], consistent with the risk assumptions associated with its new standards.

2.3 Working limits

It is essential that health professionals become aware of the underlying premises and shortcomings of the OELs that they use to assess occupational exposures. As suggested above, some limits, including many TLVs, contain little or no margins of safety because large percentages of workers exposed at or below the TLV can experience adverse health effects. Likewise, new OSHA standards, which are based on the feasibility of achieving a level in the worst segment(s) of industry, may not be sufficiently protective in sectors where exposures can be better controlled. In either case, it is prudent to adopt limits for workplace air contaminants that allow for uncertainties in the underlying risks to health and for reductions of exposures to the lowest levels feasible in

---

[4] OSHA similarly defined a STEL for benzene, stating that, "The employer shall assure that no employee is exposed to an airborne concentration of benzene in excess of ... 5 ppm as averaged over any 15 minute period" (OSHA, 1987).

[5] This quotation was taken from a particularly lucid discussion of OSHA's standard for inorganic lead that arose from legal arguments in the U.S. Court of Appeals. Interestingly, in that case, OSHA argued that the most appropriate interpretation of its PEL was that of the geometric mean exposure level (OSHA, 1991).

all segments of industry.[6] This can be done by using safety factors to modify existing OELs, based upon relevant health effects, the quality of data supporting the OEL, and the circumstances under which exposures arise (Dourson and Stara, 1983; Zielhuis and van der Kreek, 1979b; Zielhuis and van der Kreek, 1979a). Large companies in the U. S., for example, are capable of maintaining exposures at most facilities well below a new PEL and could choose a working limit of, say, one half to one fourth of the PEL on the basis of feasibility alone. Many companies have adopted corporate exposure guidelines that do this (Paustenbach and Langner, 1986). Other workplaces, which might be unable to achieve the desired levels of control, should strive for incremental reductions of average exposure levels over time (Roach, 1977). In such situations, it might initially be necessary to test exposure levels relative to the OEL. Then annual testing could be performed, with concurrent adoption of controls, until the target level of, say, one half the OEL, is achieved.

2.4 This chapter and Chapter 3

In this chapter we traced the development of OELs in the U.S. from the ACGIH TLVs to the OSHA PELs. We showed that OELs are often not what they seem, because they reflect practical issues related to workplace control rather than the health consequences of exposures. Because most health risks are related to the long-term exposures received by individual persons over a lifetime, both the evaluation and control of exposures require statistical models that can characterize levels of chemical contaminants over time and across persons in particular groups. In Chapter 3 we will delve into many of the statistical issues surrounding the collection of valid exposure data for such models.

---

[6] OSHA's promulgation of action levels (ALs) effectively imposes a working limit of one half of the PEL in many segments of industry affected by its new standards.

3 SAMPLING EXPOSURES

3.1 Establishing observational groups

The first step in assessing exposures is to assign persons to *observational groups*, based on common factors related to the personal environments. Following are some examples of observational groups: workers with the same job and/or factory, persons living in the same geographical community, suburban dwellers with garages attached to their houses, and persons living in homes with gas stoves. Observational grouping originated several decades ago with retrospective investigations of occupational diseases. Since quantitative exposure data were nonexistent (and sadly still are in most cases), investigators had to rely upon qualitative descriptors of the job and workplace as surrogates for actual exposure levels. Such qualitative classifications were originally motivated by epidemiologic studies in which it was desired to estimate ordinal levels of exposure (e.g., high, medium, and low) for groups of workers to assess associations with health outcomes.

A particularly lucid description of observational grouping for epidemiologic purposes was given by Roach (1953), who classified dust exposures of British coal miners and compared these exposures with X-ray abnormalities suggestive of simple pneumoconiosis. Somewhat later, investigators assigned various terms to designate observational groups; for example, Ashford (1958) referred to 'occupational groups', Woitowitz et al. (1970) used the term 'hazard class', and Gamble and Spirtas (1976) referred to 'occupational-title groups'. Over time, observational grouping became part of the fabric of occupational hygiene for prospective purposes, being codified first as 'exposure zones' in the late 1970s (Corn and Esmen, 1979), then as 'homogeneous exposure groups' in the 1990s (Hawkins et al., 1991), and more recently as 'similar exposure groups' (Mulhausen and Damiano, 1998).

Since observational groups are typically assigned on the basis of inspection, the grouping process is open-ended because observation can be extended to an ever-expanding array of processes, environments, and tasks. As such, grouping can be sufficiently complicated and time consuming that it can preclude monitoring of exposures *per se*. That is, available resources can be devoted so extensively to qualitative descriptors of exposure that few resources remain for quantitative exposure assessment. This appears to be the general case today, where qualitative models are used in lieu of actual measurements to assign air levels or 'control bands' (Balsat et al., 2003; Stewart and Stenzel, 1999; Swuste et al., 2003; Tait, 1992). Also, since measured exposure levels often lie outside the predicted ranges of qualitative models (Cherrie and Hughson, 2005; Jones and Nicas, 2006b; Jones and Nicas, 2006a; Tischer et al.,

2003), the consequences of relying too much on qualitative information can have adverse health consequences (Kromhout and van Tongeren, 2003).

If quantitative exposure assessment is the ultimate goal, then it is recommended in most cases that observational grouping for occupational studies be solely based on job title and location. If it is necessary to combine several job titles to include sufficient numbers of persons, grouping should generally be based upon location (e.g., room, building, or department), so that all subjects will tend to share the same environmental factors. Generically similar schemes have been applied in some studies of ambient pollutants, with observational groupings based upon either city of residence (Rappaport and Kupper, 2004) or suspected local sources [e.g., types of homes, sources of dust, school attended (Brunekreef *et al.*, 1987; Janssen *et al.*, 1999; Spengler *et al.*, 1994)].

3.1.1 Example

Suppose we wished to investigate exposures to welding fumes (a suspension of fine particles containing various metal oxides) among construction workers in U.S. industries. Figure 3.1 presents a hierarchical scheme for classifying such exposures. Within the broad class of *exposure to particulates*, we focus in this example upon *construction workers* engaged in *hot processes*, where welding fumes are generated during brazing, welding, and thermal cutting (Rappaport *et al.*, 1999; Susi *et al.*, 2000). Construction workers engaged primarily in hot processes fall into four generic job titles, defined conveniently by the particular construction trades, namely boiler makers (BM), iron workers (IW), pipe fitters (PF), and welder-fitters (WF).[7] Thus, schemes for observational grouping of construction workers exposed to welding fumes could use the construction trade and either site or type of welding process as grouping variables.

---

[7] The tasks and activities involved in these construction trades have been summarized in a set of generic job definitions. Briefly, BM (generic title 'Boiler Maker') assemble, repair or dismantle pressure vessels, tanks and vats using a variety of torches and welding equipment. IW (generic title 'Structural-Steel Worker') assemble girders, columns, etc. into large steel structures, making use of torch-cutting equipment to make alterations. PF (generic title 'Pipe Fitter') install and maintain pipe systems for steam, heating and cooling, refrigeration, etc.; this often involves precision welding of structural or stainless steel. Among members of the PF trade, it is common for certain individuals to specialize in welding procedures, in which case the work could be designated under other generic titles, particularly 'Arc Welder' or 'Welder Fitter' (WF), the latter of which is used here.

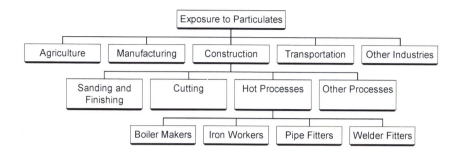

Fig. 3.1. Example of a hierarchical scheme for creating observational groups of workers exposed to welding fumes in the U.S. construction industry. The four groups shown at the bottom represent construction trades whose members are exposed to welding fumes during hot processes (brazing, welding, and thermal cutting).

3.2 Samples of data

After assigning persons to observational groups, sufficient data should be obtained to test exposures relative to limits and/or to investigate exposure-response relationships. Each measurement should be obtained by personal sampling over the full work shift[8] (for occupational exposures) or day (for environmental exposures). One or more randomly chosen exposure measurements should be acquired from each of several persons in the observational group. Sample sizes in the range of 10 - 20 measurements per observational group (i.e., two measurements from 5 - 10 randomly selected persons) should generally be sufficient for initial assessments of exposure levels (discussed in Chapter 9).

As will be seen, repeated exposure measurements for each of several persons are necessary to obtain information about the within-person and between-person variability in exposure levels in each observational group. The preferred sampling design involves *balanced data*, where the same number of measurements is obtained from each randomly chosen person (n measurements per person) in the sample. However, the ideal of balanced data is difficult to achieve in practice and statistical methods can be applied to either *balanced or unbalanced* data. In any event, sampling should be carried out over a sufficient duration of time (*one year is recommended*) to cover the full range of operational and environmental conditions (Rappaport *et al.*, 1995a; Spengler *et al.*, 1994).

---

[8] For practical purposes, the monitoring period should be at least 4 h during the work shift or day to minimize potential problems associated with combining measurements based on different averaging times.

3.3 Random versus worst-case sampling

Prior to the development of personal samplers, it was necessary to restrict the sampling effort to a handful of short-term area or breathing-zone measurements during each survey (discussed in Chapter 1). Because monitoring was technically difficult in those days, occupational hygienists attempted to identify highly exposed individuals and to ascertain whether their exposures were in the acceptable range, thereby attempting to estimate an upper bound on the exposure level for the entire group. This focus on measuring high exposure levels, referred to as *worst-case sampling*, became so deeply rooted in professional practice that it persisted after the development of personal monitors (Leidel *et al.*, 1977; Roach *et al.*, 1967), and is still encouraged (Hewett, 1997a; Hewett, 1997b). The practice probably continues because worst-case sampling is expedient within the confines of compliance testing, where all decisions hinge upon whether or not at least one measurement exceeds the OEL (Rappaport, 1984). (Compliance testing is discussed in Chapter 8).

Although worst-case sampling has some merit in the context of governmental inspections (Spear and Selvin, 1989; Tornero-Velez *et al.*, 1997), it should be resisted more generally for several reasons. First, there is little evidence that occupational hygienists can consistently identify highly exposed workers solely on the basis of observation. For example, occupational hygienists were able to predict jobs with high, medium and low exposures in some workplaces but not in others (Kromhout *et al.*, 1987; Post *et al.*, 1991). Also, Danish government inspectors who targeted worst cases reported 2-fold to 7-fold higher exposure concentrations on average than those obtained in non-targeted inspections (Olsen, 1996; Olsen *et al.*, 1991); however, the distributions of worst-case and normal exposure concentrations depended upon the contaminant being measured and were highly overlapping.

Second, selection of the worst case depends upon the particular environmental conditions or task(s) to be performed. Environmental characteristics or tasks are themselves subject to great variability because of differences in weather, ventilation, duration, process, environment, and persons being monitored (Cherrie, 1996; Nicas and Spear, 1993; Nieuwenhuijsen *et al.*, 1995; Olsen, 1994; Olsen and Jensen, 1994). This again opens the door to an ever increasing array of conditions and tasks that must be defined and characterized as potential worst cases.

Third, the intentional biased sampling of data invalidates the use of statistical tools for exposure-evaluation purposes (Olsen, 1996). This practice thus distorts inferences concerning exposure levels for the entire population and complicates statistical modeling of exposure-response relationships (Olsen, 1996; Ulfvarson, 1983) and attendant risk assessments.

3.3.1 Random sampling

All persons within an observational group should be eligible for monitoring. Each person should be assigned a number, and a particular sample of k persons for the group should be chosen randomly. If a person refuses to cooperate or leaves the group, then another individual should be randomly selected. After selecting persons, days should also be randomly sampled until n exposure measurements have been obtained from each person. If a particular person is absent on one day, then the same individual should be measured on another randomly selected occasion to complete the collection of data. It would also be useful to obtain information about environmental conditions and tasks as the measurements are made, if this can be done *relatively easily*[9]. This information can be useful at later stages to determine the impact of tasks and other factors on exposure distributions, to regroup persons, and/or to evaluate options for controlling exposure levels (discussed in Chapters 7 and 9).

If investigators can be routinely available on site, then the above recommendations about random sampling should not present major problems. However, in cases where investigators cannot be routinely on site, consideration should be given to enlisting the assistance of other professionals who are present (e.g., safety engineers, nurses, supervisory staff) or even the subjects themselves (as described below) to carry out the monitoring. After an initial period of training in the use of sampling equipment, it should be a relatively simple matter for these individuals to contribute productively to the sampling effort.

In some cases, monitoring cannot routinely be conducted on site. Then, it becomes difficult to randomly select among persons and days. In such situations, the only practical solution may well be for the investigator to make all measurements during a discrete campaign of a few days time. *Campaign sampling* can still lead to valid inferences if the full range of activities giving rise to exposure was covered during the campaign.

3.4 Self assessment of exposure

In some cases, it may be possible to maintain a rigorous plan for random sampling by relying upon the subjects to measure their own exposures. Such a strategy is particularly appealing for exposures to gaseous species where innovations in passive monitors allow workers or lay persons to easily measure their own exposures at specified times. Additional information about important covariates (e.g., activities, locations, and environmental conditions) can be obtained via questionnaires. Self assessment of exposure was first proposed by Rappaport (1991b), and has since been applied in occupational studies for the following hazardous substances: electric fields (Loomis *et al.*, 1994), benzene

---

[9] Kromhout *et al.* provide a good example of this in the rubber industry (Kromhout and Heederik, 1995; Kromhout *et al.*, 1994).

(Egeghy *et al.*, 2002; Liljelind *et al.*, 2000), styrene and terpenes (Liljelind *et al.*, 2001), and particulate matter (Rappaport *et al.*, 1999) (in this last example, trained construction workers sampled their coworkers).

Investigators in Sweden compared exposure data obtained by self assessment with data obtained more conventionally by an occupational hygienist (Liljelind *et al.*, 2001). The results, shown in Figure 3.2, suggest that the exposure distributions estimated from self assessment were very similar to those estimated by expert monitoring. Using multivariate analyses, the authors showed that the exposure levels obtained from self assessments were not statistically different from those measured by the occupational hygienist in 9 of the 10 workplaces investigated.

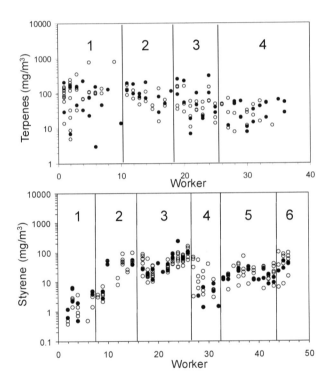

Fig. 3.2 Airborne exposures to terpenes in sawmills (top) and to styrene in reinforced plastics factories (bottom). Open circles represent self-measurements made by workers and closed circles represent measurements made by an occupational hygienist on different days. Numbers represent different workplaces of a particular type. [Data from Liljelind *et al.* (2001)].

Self assessment has also been used to measure benzene exposures for the general population during automobile refueling (Egeghy *et al.*, 2000). In that study, the investigators relied upon simple test kits to obtain measurements of benzene levels in air and breath as illustrated in Figure 3.3. Despite the simplicity of the methods, that study detected significant effects of the season of the year and type of fuel upon benzene concentrations and also showed a clear increase in levels of benzene in breath with increasing exposure levels during refueling.

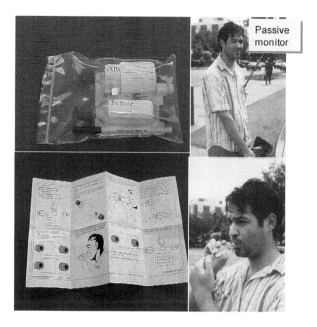

Fig. 3.3 Self assessment of benzene in air and breath during automobile refueling, as described by Egeghy *et al.* (2000). Left: Packaged air and breath samplers (top) and instructions for use (bottom). Right: Air sampling during refueling (top) and breath sampling after refueling (bottom).

These results from several investigations indicate that self assessment of exposure offers a viable, cost-effective mechanism for obtaining exposure data in both occupational and environmental populations. Hopefully, such methods will gain acceptance and enter the mainstream of practice for assessing occupational and environmental exposures in the future.

3.5 This chapter and Chapter 4

In this chapter we discussed practical issues for grouping persons according to observable factors and for obtaining sufficient numbers of measurements for statistical analyses. We emphasized the need for avoiding biases in the collection of data by randomly selecting both persons and times for monitoring. In Chapter 4, we will present the simplest statistical model that can be used to characterize exposure levels, and we will discuss the assumptions that underlie its use.

4 EXPOSURE DISTRIBUTIONS

4.1 Exposure as a random process

In order to assess characteristics of exposure levels due to environmental contaminants, it is helpful to consider exposure as a random process with numerous sources of variability. One can imagine many variables that affect the magnitudes of exposure levels at particular times. These include location, sources of contamination, type and effectiveness of controls, temperature, wind speed and direction, tasks, and work practices. The joint effects of these variables tend to produce the extraordinary ranges of exposure concentrations observed in occupational and environmental studies.

The scatter plot shown in Figure 4.1 represents a sample of occupational exposure data published three decades ago (Cope *et al.*, 1979). These 177 personal measurements (TWA levels) of inorganic lead were obtained from 6 workers in a factory producing tetraalkyllead compounds. (We will subsequently refer to this observational group as Group 1). Air levels of inorganic lead ranged from 1.3 $\mu g/m^3$ to 62.6 $\mu g/m^3$, about a 50-fold range, which is typical of continuous indoor exposure levels in manufacturing industries (Kromhout *et al.*, 1993). Also note that a few of the daily measurements were above the OSHA PEL of 50 $\mu g/m^3$.

When formed into a histogram, as depicted in the top section of Figure 4.2, the air concentrations of lead show the characteristic right skewness of occupational and environmental measurements, with generally small values punctuated by a few very large observations. Since such skewness is a property of the lognormal distribution, it is now common practice to treat occupational exposure data as lognormally distributed for statistical purposes (Esmen and Hammad, 1977; Oldham, 1953; Rappaport, 1991b). If air concentrations are lognormally distributed, then the logged values would be normally distributed and, therefore, would be amenable to analysis using statistical methods where an assumption of at least approximate normality is needed. Referring again to the airborne lead concentrations, after taking natural logarithms, the histogram assumes an approximate bell-shaped curve as shown in the bottom section of Figure 4.2. This figure can be compared with the data representing coal dust exposures reproduced from Oldham's 1953 paper, shown as Figure 4.3 (Oldham, 1953).

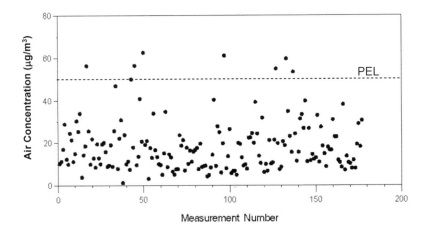

Fig. 4.1 Scatter plot of air exposure concentrations to inorganic lead ($N = 177$, $k = 6$) for a group of workers in a tetraalkyllead manufacturing plant (Group 1) [Data from Cope *et al.* (1979)].

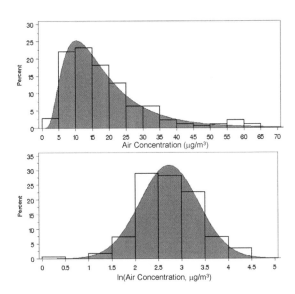

Fig. 4.2 Histograms of 177 personal measurements of inorganic lead air concentrations depicted in Figure 4.1. Top: Exposure distribution showing characteristic right skewness (the shaded area represents the best fit of a lognormal distribution); Bottom: Distribution of the natural logarithms of the exposure concentrations (the shaded area represents the best fit of a normal distribution).

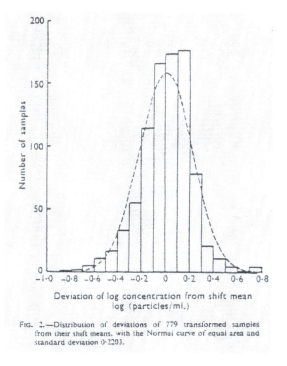

FIG. 2.—Distribution of deviations of 779 transformed samples from their shift means, with the Normal curve of equal area and standard deviation 0·2203.

Fig. 4.3 Histogram of logged deviations of 779 breathing-zone measurements of dust in British coal mines [from Oldham (1953)]. (*Brit J Ind Med*, 1953; 10: 227-234; reproduced with permission from the BMJ Publishing Group).

4.2 Lognormal distributions of exposures

In order to use relatively small numbers of measurements to make reasonable inferences about the parameters characterizing underlying distributions of exposures, it is necessary to use valid parametric statistical models (i.e., models that involve specific assumptions like normality, homogeneous variance, etc.). The first statistical models for assessing exposures in observational groups (based upon area or breathing-zone samples) assumed that all group members experienced the same average air levels over time. This view of exposure, and the realization that air levels changed from day to day and were approximately lognormally distributed, gave rise to the following model of exposure (Ashford, 1958; Breslin *et al.*, 1967; LeClare *et al.*, 1969; Sherwood, 1966; Sherwood, 1971; Tomlinson, 1957):

$$Y_j = \ln(X_j) = \mu_Y + e_j, \qquad \text{for } j = 1, 2, ..., N \text{ daily exposures,} \qquad (4.1)$$

where X_j represents the jth daily exposure concentration received by all members of the observational group, Y_j is the natural logarithm of X_j, μ_Y

represents the fixed true mean (logged) exposure level, and e_j is the error term representing all sources of random variation including random assay errors. It is assumed under Model (4.1) that the $\{e_j\}$ are mutually independent and each normally distributed with mean zero and variance σ_Y^2, so that the $\{Y_j\}$ are necessarily normally distributed with mean μ_Y and variance σ_Y^2. Thus, the $\{X_j\}$ are necessarily lognormally distributed with mean $\mu_X = e^{(\mu_Y + 0.5\sigma_Y^2)}$ and variance $\sigma_X^2 = \mu_X^2 \left(e^{\sigma_Y^2} - 1 \right)$ (Aitchison and Brown, 1957). Thus, fixing on the jth day, the average exposure level on that day is $e^{(\mu_Y + e_j)}$ under Model (4.1).

Model (4.1) would be appropriate for characterizing exposure levels in two situations. First, there could be instances where one measurement is obtained on each of several randomly selected persons in an observational group, so that these measurements can be assumed to be mutually independent of one another. In this case, μ_X and μ_Y represent the true means of the exposure levels and logged exposure levels, respectively, for this observational group. Second, there could be situations where repeated measurements are obtained on one person and it can be assumed that these measurements are not correlated with (i.e., are mutually independent of) one another. In this case, μ_X and μ_Y would represent the true means of the exposure level and logged exposure level for that person. The latter situation is illustrated in Figure 4.4 for the 36 air measurements of inorganic lead from the first worker in Group 1. For both of these situations, it is assumed that μ_X (and hence μ_Y) and σ_X^2 (and hence σ_Y^2) do not change over the time frame during which the exposure measurements are obtained (see Section 4.4).

Under Model (4.1), various parameters characterizing the lognormal density of the random variable $X_j = e^{Y_j}$ (e.g., μ_X and σ_X^2) can be directly related to the corresponding parameters of the normal distribution (e.g., μ_Y and σ_Y^2) of the random variable Y_j. In particular, the geometric mean μ_{gX} of X_j is directly related to μ_Y via the equation $\mu_{gX} = e^{(\mu_Y)}$, and the geometric standard deviation of X_j is defined as $\sigma_{gX} = e^{(\sigma_Y)}$. Table 4.1 provides useful relationships involving μ_Y, σ_Y, μ_X, σ_X, μ_{gX}, and σ_{gX}.

Exposure Distributions

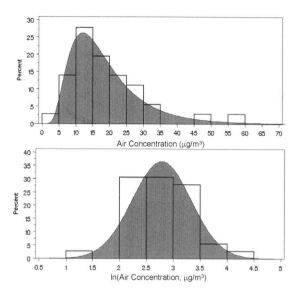

Fig. 4.4 Histograms of air levels (top) and logged air levels (bottom) measured for Subject No. 1 in Group 1. [Shaded areas represent the best fits to the data of a lognormal distribution (top) and a normal distribution (bottom)].

Table 4.1 Formulas for converting among parameters of lognormal and normal distributions. [Adapted from Leidel *et al.* (1977)].

Given	To Obtain	Use
μ_Y	$\mu_{gX} =$	$\exp(\mu_Y)$
μ_X, σ_X	$\mu_{gX} =$	$\mu_X^2 / \sqrt{\mu_X^2 + \sigma_X^2}$
σ_Y	$\sigma_{gX} =$	$\exp(\sigma_Y)$
μ_X, σ_X	$\sigma_{gX} =$	$\exp\sqrt{\ln[1+(\sigma_X^2/\mu_X^2)]}$
μ_Y, σ_Y	$\mu_X =$	$\exp(\mu_Y + \sigma_Y^2/2)$
μ_{gX}, σ_Y	$\mu_X =$	$\mu_{gX}[\exp(\sigma_Y^2/2)]$
μ_Y, σ_Y	$\sigma_X =$	$\sqrt{[\exp(2\mu_Y+\sigma_Y^2)][\exp(\sigma_Y^2)-1]}$
μ_{gX}, σ_Y	$\sigma_X =$	$\sqrt{\mu_{gX}^2[\exp(\sigma_Y^2)][\exp(\sigma_Y^2)-1]}$
μ_{gX}	$\mu_Y =$	$\ln(\mu_{gX})$
μ_X, σ_Y	$\mu_Y =$	$\ln(\mu_X) - \sigma_Y^2/2$
σ_{gX}	$\sigma_Y =$	$\ln(\sigma_{gX})$
μ_X, σ_X	$\sigma_Y =$	$\sqrt{\ln[1+(\sigma_X^2/\mu_X^2)]}$

4.3 Stationarity

The term *stationarity* implies that the statistical parameters of the underlying distribution of exposure concentrations, particularly the mean and variance, do not change over the time period of interest. The work of Symanski and coworkers offers some insight into the stationarity of occupational exposures (Symanski *et al.*, 1998a; Symanski *et al.*, 1996; Symanski *et al.*, 1998b; Symanski and Rappaport, 1994). Regarding long-term exposures (years to decades), the authors identified pervasive trends, generally toward lower exposure levels (observed in 78% of 696 datasets), with rates of reduction ranging between 4% and 14% per year (median 8% reduction per year) (Symanski *et al.*, 1998a; Symanski *et al.*, 1998b). Some of these trends are shown in Figure 4.5, where surprisingly strong linear relationships were observed between the log-transformed average exposure levels and time. Several factors were found to influence the rate of reduction in average exposure levels, including the industrial sector and the type of contaminant (gas or vapor). On the other hand, relatively little change in average exposure levels was detected in surveys conducted less than one year apart (Symanski *et al.*, 1996), and most time series of consecutive daily measurements obtained from the same workers within one month appeared to be stationary (Symanski and Rappaport, 1994).

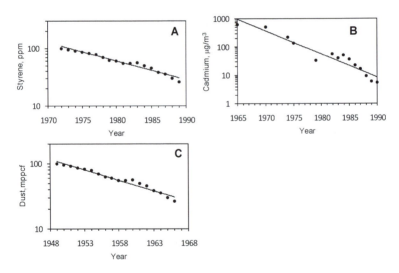

Fig. 4.5 Examples of long-term trends in occupational exposure to chemicals, compiled by Symanski *et al.* (1998a). A: Styrene exposure in a reinforced-plastics factory (Norway); B: Cadmium exposure in a battery manufacturing plant (U.K.); C: Dust exposure in an asbestos mine (Canada).

Based upon these results, exposure assessment should be viewed as an ongoing activity in which inferences about the levels of airborne chemicals are made periodically. Since most trends appear to be toward lower exposure levels, a finding of low exposure during one period (e.g., year) would suggest that even lower levels might follow.

4.4 Autocorrelated exposure series

Under Model (4.1), the set of logged exposure levels received by a person at different times is given by the $\{Y_j\}$, which are assumed to be mutually independent and normally distributed with mean μ_Y and variance σ_Y^2. However, this model ignores the possible correlation among exposures measured at different times. It is assumed under Model (4.1) that the correlation between any pair of Y_j values is equal to zero, regardless of the fixed time difference (or time lag) between measurements. If the pairwise correlation was, in fact, related to the time lag, then this assumption of mutual independence would be incorrect and ignoring such *autocorrelation* could lead to biased estimates of the parameters of the exposure distribution (Francis *et al.*, 1989; Symanski *et al.*, 1996). Such bias could be introduced, for example, through the practice of *campaign sampling*, where all measurements are obtained during a few consecutive days.

To model the temporal correlation of exposure levels, an *autocorrelation function* is needed to describe the correlation between exposure measurements separated by a fixed interval of time, referred to as the lag. Given the importance of minimizing the effects of bias when estimating parameters of an exposure distribution, it would be useful to accurately quantify the structure of the appropriate autocorrelation function for exposure measurements. Unfortunately, a reasonable estimate of the autocorrelation function over the first 10 - 12 sequential (equally spaced) time lags requires measurement of at least 50 sequential exposures (Box and Jenkins, 1976). Since such large numbers of personal measurements are rarely collected in practice, relatively little information is currently available to allow autocorrelation functions to be properly estimated. This paucity of useful data has motivated investigators to rely upon simpler models to gain insight into the potential effects of autocorrelation on the estimation of parameters of exposure distributions. The most popular of these simpler models has been the first-order autoregressive process [AR(1) process], which depicts the current exposure as a fraction of the previous exposure plus a random input (Coenen, 1976; Francis *et al.*, 1989; George *et al.*, 1995; Koizumi, 1980; Preat, 1987; Rappaport and Spear, 1988; Roach, 1977; Spear *et al.*, 1986; Symanski and Rappaport, 1994).

For an AR(1) process, we can consider modeling the correlation between the jth and $(j+1)^{th}$ observations from a time series of equally spaced exposure concentrations using the following model:

$$(Y_{j+1} - \mu_Y) = \alpha(Y_j - \mu_Y) + e_j, \qquad \text{for } j = 1, 2, ..., N \text{ exposure levels;} \quad (4.2)$$

here, $\alpha = \text{Corr}(Y_j, Y_{j+1})$ represents the autocorrelation (weighting) parameter. If we define q as the number of equally spaced time lags between any pair of exposure measurements Y_j and $Y_{j'}$, then the autocorrelation function for the AR(1) process is given by $\rho(q) = \alpha^q$. Since α tends have a value between zero and one for occupational exposures, the autocorrelation function $\rho(q)$ thus tends to decline with increasing lags between pairs of exposure levels in a given workplace. That is, as the time between any pair of exposure measurements increases, the correlation between the two measurements decreases.

In the context of day-to-day exposures, there are only anecdotal suggestions that significant autocorrelation exists. For example, Ulfvarson (1983) commented upon seasonal variations, and Buringh and Lanting (1991) observed that the variances of small-sized data sets, obtained from a variety of industries, were smaller when all measurements were collected within a single week than otherwise. However, in the only studies to investigate the question directly (with time series of daily exposures), relatively little evidence of autocorrelation was observed; and, when found, the correlation coefficient representing sequential values (termed the 'first-lag' coefficient) tended to be rather small (less than about 0.3) (Francis *et al.*, 1989; George *et al.*, 1995; Kumagai *et al.*, 1993; Symanski and Rappaport, 1994). If these results are typical of the negligible day-to-day correlation observed in most workplaces, then it should be possible to validly estimate the important parameters of distributions of daily measurements of exposure on the basis of sampling campaigns of a few days duration. However, as noted earlier, potential problems can be avoided by random sampling.

Some work has been published concerning the levels of autocorrelation of intraday exposure concentrations. Coenen (1971; 1976) noted that dust concentrations and vinyl chloride concentrations measured continuously at fixed locations in manufacturing facilities were highly autocorrelated on a given day. Kumagai *et al.* (1993) reported a similar finding for short-term personal exposure samples obtained from 16 worker-chemical combinations, and they also presented evidence that the AR(1) process was appropriate for such time series. Roach (1977), Spear *et al.* (1986), and Rappaport and Spear (1988) used an AR(1) process, based upon air exchange rates, to predict short-term autocorrelations in occupational environments (where mass transport of the contaminant is governed by turbulent diffusion). These models suggest that short-term exposures would be highly autocorrelated under realistic scenarios. Collectively, these studies indicate that periods of at least a few hours between measurements can be required to obtain relatively uncorrelated data. Thus, the occupational hygienist wishing to predict the frequencies of brief excursions above short-term-exposure limits (STELs) should be wary of doing so on the basis of measurements taken close together in time (discussed further in Chapter 8, Section 8.6.2). Recent advances in the development of personal monitors that measure short-term exposures, and that store the data over an entire shift, allow intraday autocorrelation functions to be investigated.

4.5 Measurements below the limit of detection

Exposure levels are frequently reported as being less than the analytical limit of detection (LOD) for a particular method of measurement. Such unobservable levels are referred to as (left) censored values. It is not uncommon to encounter situations where more than 10% of the exposure measurements are left-censored. For example, among 44 benzene-exposed groups in the petroleum-refining industry, Spear et al. (1987) found between 0 and 64% left-censoring, with a median value of 12%. Replacing censored values with fixed values generally leads to biased estimates of the mean and variance of the underlying exposure distribution; e.g., substitution of the LOD for censored values overestimates the mean exposure. Hornung and Reed (1990) showed that replacing censored values with either LOD/$\sqrt{2}$ (when $\sigma_{gX} \leq 3$) or LOD/2 (when $\sigma_{gX} > 3$) minimized this estimation bias for lognormally distributed exposure data with minor censoring (less than 10 – 20%). With more extensive censoring, statistical methods are required to properly analyze a data set containing left-censored observations. These methods range from rather simple graphical procedures (Travis and Land, 1990) to complex iterative and maximum-likelihood methods (Dempster et al., 1977; Hughes, 1999; Lange, 1999; Taylor et al., 2001)[10].

4.6 Estimation of parameters

As noted above, it is assumed under Model (4.1) that the $\{Y_j\}$, representing the set of logged exposures from a particular person or group, are mutually independent and normally distributed random variables, each with mean μ_Y and variance σ_Y^2. The parameters μ_Y and σ_Y^2 can be estimated using the estimators $\hat{\mu}_Y = \frac{1}{N}\sum_{j=1}^{N} Y_j$ and $\hat{\sigma}_Y^2 = \frac{1}{N-1}\sum_{j=1}^{N}(Y_j - \hat{\mu}_Y)^2$, where the '^s' designate $\hat{\mu}_Y$ and $\hat{\sigma}_Y^2$ as estimators of μ_Y and σ_Y^2, respectively. These are minimum variance unbiased estimators (MVUEs) of μ_Y and σ_Y^2, with variances given by $V(\hat{\mu}_Y) = \sigma_{\hat{\mu}_Y}^2 = \frac{\sigma_Y^2}{N}$ and $V(\hat{\sigma}_Y^2) = \frac{2\sigma_Y^4}{N-1}$, respectively (Aitchison and Brown, 1957).

Regarding the lognormal distribution of exposures $\{X_j\}$ for this person or observational group, the geometric mean and geometric standard deviation μ_{gX} and σ_{gX} can be directly estimated as $\hat{\mu}_{gX} = \exp(\hat{\mu}_Y)$ and $\hat{\sigma}_{gX} = \exp(\hat{\sigma}_Y)$. The mean and variance of the lognormal distribution, i.e., μ_X and σ_X^2, are estimated

---

[10] Note that Hughes (1999) dealt with situations where censored data included repeated observations from multiple subjects that were analyzed with mixed effects models (discussed in Chapter 6).

as $\hat{\mu}_X = \exp\left(\hat{\mu}_Y + \frac{\hat{\sigma}_Y^2}{2}\right)$ and $\hat{\sigma}_X^2 = \exp\left(2\hat{\mu}_Y + \hat{\sigma}_Y^2\right)\left(\exp(\hat{\sigma}_Y^2) - 1\right)$. These estimators of μ_X and σ_X^2 have properties that make them preferable to the simple unbiased estimators $\bar{X} = \frac{1}{N}\sum_{j=1}^{N} X_j$ and $S_X^2 = \frac{1}{N-1}\sum_{j=1}^{N}(X_j - \bar{X})^2$ for lognormally distributed data. As noted by Oldham (1953), although $\bar{X}$ and S_X^2 are unbiased estimators of μ_X and σ_X^2, they have larger variances (i.e., are less precise) than the estimators $\hat{\mu}_X$ and $\hat{\sigma}_X^2$. The estimators $\hat{\mu}_X$ and $\hat{\sigma}_X^2$ are slightly biased, with the bias being negligible if N is moderately large.

For each of the 6 workers in Group 1 (exposed to inorganic lead), the estimated parameters for the normal and lognormal distributions under Model (4.1) are presented in Table 4.2. For these data, $\hat{\mu}_Y$, $\hat{\sigma}_Y^2$, $\bar{x}$, and s_X^2 were computed as shown above, while $\hat{\mu}_{gX}$, $\hat{\sigma}_{gX}$, $\hat{\mu}_X$, and $\hat{\sigma}_X^2$ were computed by substituting $\hat{\mu}_Y$ and $\hat{\sigma}_Y^2$ for μ_Y and σ_Y^2 in the relationships given in Table 4.1. Note that the estimated mean of the lognormal distribution, $\hat{\mu}_X$, is larger than that of the corresponding estimated geometric mean, $\hat{\mu}_{gX}$, and that the simple estimates of the mean and variance of the lognormal distribution, $\bar{x}$ and s_X^2, are generally close to the values of $\hat{\mu}_X$ and $\hat{\sigma}_X^2$. The single exception occurs for Subject 2, where $\hat{\sigma}_X^2 = 694.5$ is much larger than $s_X^2 = 274.7$.

Table 4.2 Estimates of the parameters of the normal and lognormal distributions of exposure levels (μg/m³) for each of 6 workers exposed to inorganic lead (Group 1). Estimates were obtained under Model (4.1).

Subject	N	$\hat{\mu}_Y$	$\hat{\sigma}_Y^2$	$\hat{\mu}_{gX}$	$\hat{\sigma}_{gX}$	$\hat{\mu}_X$	$\bar{x}$	$\hat{\sigma}_X^2$	s_X^2
1	36	2.78	0.298	16.2	1.73	18.8	18.7	122.4	118.9
2	23	2.79	0.788	16.2	2.43	24.1	22.0	694.5	274.7
3	34	2.47	0.315	11.8	1.75	13.8	13.8	71.1	70.4
4	36	2.67	0.425	14.5	1.92	17.9	17.9	169.6	167.5
5	33	3.01	0.238	20.2	1.63	22.8	22.8	139.6	144.9
6	15	2.62	0.275	13.8	1.69	15.8	15.8	79.0	90.6

4.7 Normality of (logged) individual exposure levels

The estimated parameters of the individual exposure distributions shown in Table 4.2 provide little evidence as to whether the $\{Y_j\}$ are approximately normally distributed, as assumed for Model (4.1). The fact that the estimated parameters of the lognormal distributions, $\hat{\mu}_X$ and $\hat{\sigma}_X^2$, obtained by substituting $\hat{\mu}_Y$ and $\hat{\sigma}_Y^2$ into the formulae in Table 4.1, were similar to the simple estimates $\bar{x}$ and s_x^2 provides indirect evidence that the underlying exposure distributions of the $\{X_j\}$ are approximately lognormal (and, by extension, that the $\{Y_j\}$ are approximately normally distributed).

To investigate the normality assumption directly, normal probability plots are shown in Figure 4.6 for the 6 individual exposure distributions from Group 1. If the normal distribution percentiles, plotted against the observed cumulative distributions of logged measurements from a given person, fall approximately along a straight line, then it is reasonable to infer that the $\{Y_j\}$ are roughly normally distributed (Aitchison and Brown, 1957). As shown in Figure 4.6, the observed $\{y_j\}$ from Group 1 generally fall along straight lines, even though the data from Subjects 2 and 6 display some discordant observations. A useful test for goodness of fit of the normal distribution is the Shapiro-Wilks W test, which has been applied to occupational exposure data in a few cases (Kumagai et al., 1997; Kumagai and Matsunaga, 1995; Waters et al., 1991). Results from application of the W test to the 6 individual (logged exposure) data sets from Group 1 are summarized in Table 4.2. In each case, the P-value is greater than 0.05, indicating that there is not strong statistical evidence to reject the null hypothesis of a normal distribution (of logged exposure measurements). However, note that the data sets for Subjects 2 and 6 are associated with P-values of about 0.1, suggesting marginal lack of fit to normal distributions for these two persons.

4.8 This chapter and Chapter 5

In this chapter, we showed that concentrations of air contaminants tend to cover large ranges for a given group of persons and that the distributions of such levels exhibit characteristic (right) skewness toward large values. Because these features are properties of lognormal distributions, we introduced Model (4.1) to summarize the important characteristics of lognormal distributions of exposure concentrations via a few parameters. We then showed how these parameters can be related to distribution parameters for corresponding normal distributions of logged exposure levels. Because Model (4.1) requires that all measurements be mutually independent, it cannot be applied in common situations where multiple (correlated) exposure measurements are obtained from each of several randomly selected group members; therefore, Model (4.1) cannot generally be used to characterize the variability of exposure concentrations both within persons and between persons in the same group. In Chapter 5, we will introduce a more sophisticated statistical model that is useful for evaluating variability of exposure concentrations within and between group members.

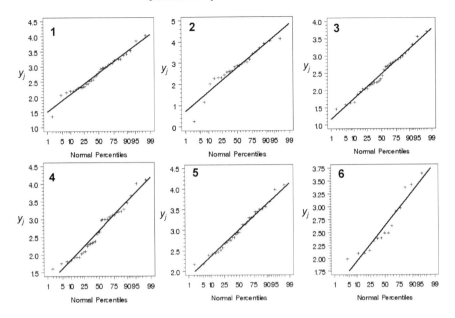

Fig. 4.6. Normal probability plots for the sets of logged air measurements, $\{y_j\}$, obtained from the 6 workers (identified by number) in Group 1.

Table 4.3 Values of the Shapiro-Wilks W-Statistic and associated P-values for the sets of individual (logged) measurements for the 6 workers in Group 1.

Subject	N	W-Statistic	P-Value
1	36	0.9801	0.7496
2	23	0.9298	0.1086
3	34	0.9763	0.6523
4	36	0.9539	0.1385
5	33	0.9778	0.7169
6	15	0.9033	0.1070

5 EXPOSURE VARIABILITY WITHIN AND BETWEEN PERSONS

The scatter plot of personal exposures to inorganic lead illustrated in Figure 4.1 is difficult to interpret because it was compiled from repeated measurements from 6 different workers in Group 1. The estimated day-to-day variances of air lead levels, which were investigated via Model (4.1) in Chapter 4, indicated that levels varied greatly *within persons*. However, since Model (4.1) could not validly be applied to all the data from Group 1, it provided no insight into how mean exposure levels might have varied *between persons* in that group. Certainly, the estimated mean exposure levels (values of $\hat{\mu}_X$) for the 6 workers from Group 1, ranging between 13.8 and 24.1 µg/m^3 (see Table 4.2), suggest that differences in mean exposures were rather small for this group. But, in general, it would be useful to be able to estimate both within-person and between-person sources of variation in an observational group using a single statistical model of exposure. Such models are generally referred to as *random-effects models* or *analysis of variance (ANOVA) models*.

5.1 Importance of between-person variability

Within-person and between-person sources of variability in exposure levels were recognized as early as 1952 when Oldham and Roach applied ANOVA models to breathing zone samples of dust in British coal mines (Oldham and Roach, 1952). They made the following observation:

> "It was found that significant variation was occurring in the dust concentrations from one collier's experience to another's, and from one day to another in the same collier's experience." (Note that a 'collier' is a coal miner).

Yet, this finding was largely ignored at the time, and the issue of within-person and between-person variability was not revisited again until some 35 years later when personal sampling became available in occupational studies (Kromhout *et al.*, 1987; Rappaport *et al.*, 1988b; Spear *et al.*, 1987). Data collected with personal monitors in the 1970s and 1980s provided strong evidence that considerable heterogeneity in mean exposure levels existed between workers having the same job at a particular factory. This is a reasonable finding because, even though persons may have the same job, they can work in different locations, perform different tasks, use different types of equipment, and have different levels of training and experience. Likewise, persons living in a given community can encounter different indoor and outdoor sources of air pollution, can reside in houses with different levels of air exchange, can be exposed to different levels of active and/or passive cigarette smoke, etc.

Regardless of the reasons why persons in an observational group are exposed to different concentrations of a given contaminant, the concept of between-person variability is central to the assessment of exposure for both hazard control and epidemiologic research (Rappaport, 1991b). If all persons in a group were exposed to exactly the same mean contaminant level, then there would be no between-person variability and the group would be *uniformly exposed*[11] (Rappaport, 1991b; Rappaport et al., 1993). In such cases, all group members would be at roughly equal health risk due to the effects of long-term exposure, the same average level of exposure could be assigned to all persons for an epidemiologic study, and group-level interventions (engineering or administrative controls, which affect all members more-or-less equally) would be sufficient to reduce air levels uniformly for all workers. However, if substantial variation in mean exposure levels exists between persons in an observational group, the distribution of exposure-related risk across the population would be heterogeneous. As a consequence, exposure-control options would change from the group level to the individual level (by focusing upon personal environments), and apparent relationships between exposure levels and health outcomes would be misleading due to assignment of the same mean exposure level to all persons in the group. (These issues are considered in Chapters 9 and 10).

5.2 One way random effects model

In order to accommodate random variation in mean exposure levels between persons in an observational group, it is necessary to add a between-person random effect to Model (4.1). This addition leads to the *one way random effects model* defined as:

$$Y_{ij} = \ln(X_{ij}) = \mu_Y + b_i + e_{ij} \qquad (5.1)$$
for $i = 1, 2, ..., k$ persons and $j = 1, 2, ..., n_i$ days;

here, X_{ij} represents the exposure level for the ith person on the jth day. Following our previous convention, we designate the mean and variance of X_{ij} as μ_X and σ_X^2, respectively, and those of Y_{ij} as μ_Y and σ_Y^2, respectively. Under Model (5.1), μ_Y represents the true fixed mean (logged) exposure level for the group; b_i represents the random effect for the ith person [given by the deviation of the ith person's true mean (logged) exposure level from μ_Y, i.e.,

---

[11] This designation of uniform exposure was developed to avoid confusion with other terms, such as the "homogeneous exposure group", which can, in fact, represent highly heterogeneous exposures among the members of a group (Rappaport, 1991b; Rappaport et al., 1993).

$b_i = \mu_{Y_i} - \mu_Y$]; and, e_{ij} represents the random deviation of the observed logged exposure level Y_{ij} on the jth day for person i from μ_{Y_i} (i.e., $e_{ij} = Y_{ij} - \mu_{Y_i}$). Following the early application of Model (5.1) by Oldham and Roach (1952), this model was subsequently applied around 1990 to evaluate occupational exposures (Kromhout *et al.*, 1987; Kromhout *et al.*, 1993; Rappaport, 1991b; Spear *et al.*, 1987), and it has since been used to model exposure data for hundreds of observational groups.

Here, we will apply Model (5.1) to data from Group 1, as well as to data from three other observational groups designated as Groups 2 - 4 (see Table 5.1). In addition to the measurements of inorganic lead (Group 1), these data include air levels of benzene from the petroleum-refining industry (Groups 2 and 3) [data from Spear *et al.* (1987)], and air levels of styrene in the reinforced-plastics industry (Group 4) [data from Rappaport *et al.* (1996)]. Figure 5.1 provides plots of all (natural) logged air concentrations (values of $\{y_{ij}\}$) for Groups 1 - 4 and also gives the estimated mean value of the logged measurements for each worker (designated with a '+'). For each group, individual air levels varied greatly within persons from day to day, consistent with fold ranges (highest value/lowest value) between 48.2 and 12,400, as shown in Table 5.1. However, between-person variability in mean logged air levels was reasonably small for Groups 1 and 2, moderate for Group 4, and large for Group 3.

Table 5.1 Descriptions of data sets from four observational groups of workers.

Group	Chemical	Industry	Job	N	k	Fold-Range
1	Inorganic lead	Alkyllead manufacturing	Unknown	177	6	48.2
2	Benzene	Petroleum refining	Refining operator	90	18	63.9
3	Benzene	Petroleum refining	Transfer operator	48	24	12,400
4	Styrene	Boat manufacturing	Sprayer or laminator	103	19	141

5.2.1 Assumptions

Under Model (5.1), it is assumed that the b_i and e_{ij} are mutually independent and normally distributed random variables, with means of zero and variances σ_{bY}^2 and σ_{wY}^2, representing between-person and within-person variability, respectively. Thus, the total variability in logged exposure levels experienced by a group is given by $\sigma_Y^2 = \sigma_{bY}^2 + \sigma_{wY}^2$. Also, under Model (5.1), $Y_{ij} = \ln(X_{ij})$ is normally distributed with mean μ_Y and variance σ_Y^2.

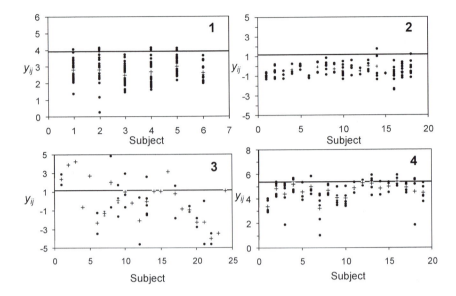

Fig. 5.1. Scatter plots of logged exposure concentrations for four observational groups of workers described in Table 5.1 (shown by group number). Each point (y_{ij}) represents the daily log-transformed air concentration measurement ($\mu g/m^3$ for Group 1 or mg/m^3 for Groups 2 – 4) for the ith subject on the jth day, and each (+) represents the estimated mean (logged) exposure concentration for that subject. Horizontal lines represent the logged Permissible Exposure Limits (PELs) at the time of monitoring.

The following assumptions are also implicit in Model (5.1):

1. The expected value of Y_{ij} is μ_Y for all i and j; and, the expected value of Y_{ij} given b_i fixed is $\mu_{Y_i} = (\mu_Y + b_i)$, indicating that the expected value of each Y_{ij} conditional on (i.e., fixing on) the ith person is equal to μ_{Y_i}.

2. The variance of Y_{ij} given b_i (or equivalently μ_{Y_i}) fixed, namely $V(Y_{ij}|\mu_{Y_i})$, equals σ^2_{wY} (indicating that daily exposures for person i vary about μ_{Y_i} with variance σ^2_{wY}). Thus, given b_i or μ_{Y_i} fixed, Y_{ij} is normally distributed with mean $\mu_{Y_i} = (\mu_Y + b_i)$ and variance σ^2_{wY}. We empirically investigated the distribution of $\{y_{ij}\}$ values for each individual worker from Group 1 in Chapter 4.

3. The within-person variance component σ^2_{wY} does not vary with i (i.e., there is an assumption of *homogeneous variance,* so that the intraperson variability is the same for all individuals in the group).

4. For $j \neq j'$, the covariance between Y_{ij} and $Y_{ij'}$ (a pair of exposure measurements for the ith person collected at different times) is equal to σ_{bY}^2, so that the correlation Corr(Y_{ij}, $Y_{ij'}$) between Y_{ij} and $Y_{ij'}$ is $\rho = \sigma_{bY}^2 / \sigma_Y^2$ (this is the so-called intraclass correlation where the class here is the individual person). This means that only positive correlation is allowed between pairs of exposure levels from the same person and that this positive correlation is the same for any pair of exposure levels, regardless of how far apart they are spaced in time (i.e., the exposures are assumed not to be autocorrelated; see Section 4.5).

These assumptions have been evaluated and tested with databases of occupational exposures (Kromhout et al., 1993; Lyles et al., 1997b; Rappaport et al., 1995a; Tornero-Velez et al., 1997). The use of graphical procedures to evaluate the normality assumption of the between-person random effect b_i (Tornero-Velez et al., 1997) indicated that Model (5.1) adequately fit data from 220 of 252 (87%) observational groups. This strongly suggests that Model (5.1) is generally appropriate for applications involving occupational exposure data.

Figure 5.2 illustrates various hypothetical probability density functions (PDFs) under Model (5.1). The PDF labeled 'Group' in Figure 5.2A refers to the distribution of logged exposure concentrations for a hypothetical group where Y_{ij} is normally distributed with mean $\mu_Y = 2.3$ and variance $\sigma_Y^2 = 0.693$. It is assumed that the within-person and between-person components of variance are equal for this group, namely, $\sigma_{wY}^2 = \sigma_{bY}^2 = 0.693/2 = 0.346$. The numbered PDFs in Figure 5.2A illustrate the distributions of Y_{ij} given b_i (or μ_{Y_i}) for a random sample of 5 persons ($i = 1, 2, 3, 4, 5$), with the ith density being normal with mean $\mu_{Y_i} = (\mu_Y + b_i)$ and variance σ_{wY}^2. Note that only the Y_{ij}'s, and not the μ_{Y_i}'s, are observable. Also, under Model (5.1), the μ_{Y_i}'s for $i = 1, 2, 3, 4, 5$ are assumed to constitute a random sample of size $k = 5$ from a normal PDF with mean μ_Y and variance σ_{bY}^2 (Figure 5.2B).

5.2.2 Lognormal distributions

Although Model (5.1) is applied to the logged exposure levels represented by the set of normally distributed $\{Y_{ij}\}$, it provides a valuable tool for making inferences about the underlying lognormal distribution representing the set of actual exposure levels, i.e., the $\{X_{ij}\} = \{X_{11}, X_{12}, ..., X_{kn_k}\}$. This is because it is implicit in Model (5.1) that $X_{ij} = e^{(\mu_Y + b_i + e_{ij})}$ is lognormally distributed with mean $\mu_X = e^{(\mu_Y + 0.5\sigma_Y^2)}$ and variance $\sigma_X^2 = \mu_X^2(e^{\sigma_Y^2} - 1)$, in agreement with most current knowledge of the distributions of occupational exposures (Rappaport, 1991b). Therefore, even though Model (5.1) pertains to the logged values $\{Y_{ij}\}$, it also establishes a mechanism for directly relating the corresponding $\{X_{ij}\}$ to internal dose and with the subsequent risk of disease (Rappaport, 1991b; Rappaport, 1993a) (discussed in Chapter 11).

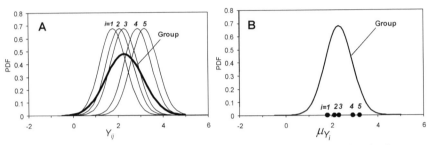

Fig. 5.2 Hypothetical normal distributions of logged exposure levels for an observational group. A) Distributions of logged exposure levels (Y_{ij}) received from day to day. Normal PDF's labelled 1 – 5 represent a random sample of 5 persons from the group (each with mean μ_{Y_i} and variance σ_{wY}^2), while the curve labelled 'Group' represents the hypothetical normal PDF (with mean μ_Y and variance σ_Y^2) of all daily exposures for the group. B) Points represent the individual mean (logged) exposure levels for the sample of 5 persons depicted in A. The PDF labelled 'Group' represents the normal PDF of individual mean (logged) exposures μ_{Y_i}, with mean μ_Y and variance σ_{bY}^2, for the group.

Since $Y_{ij} = \mu_Y + b_i + e_{ij}$, then $X_{ij} = e^{(\mu_Y + b_i + e_{ij})}$, so that E($X_{ij}$|subject i) = μ_{X_i} = $e^{(\mu_Y + b_i + 0.5\sigma_{wY}^2)}$ and V(X_{ij}|subject i) = $\sigma_{X_i}^2 = \mu_{X_i}^2 (e^{\sigma_{wY}^2} - 1)$. Equivalently, the set of n_i exposure levels $\{X_{ij}\}$ for subject i constitutes a random sample of size n_i from a lognormal density with mean μ_{X_i} and variance $\sigma_{X_i}^2$. Here, $\sigma_{X_i}^2$ is the parameter reflecting within-person variability for subject i. Figure 5.3A illustrates the conditional distributions of exposure levels for the 5 hypothetical subjects considered in Figure 5.2. Note that $\sigma_{X_i}^2 = V(X_{ij}|$subject i) = $\mu_{X_i}^2(e^{\sigma_{wY}^2} - 1)$ increases as μ_{X_i} increases. The curve labeled 'Group' again refers to the hypothetical exposure distribution for the group as a whole, just as it did for the log-transformed values in Figure 5.2A.

It is also implicit in Model (5.1) that the unobservable set of k mean exposures $\{\mu_{X_i}\} = \{\mu_{X_1}, \mu_{X_2}, ..., \mu_{X_k}\}$ constitutes a random sample of size k from a lognormal distribution with overall (group) mean $\mu_X = E(\mu_{X_i}) = e^{[\mu_Y + 0.5(\sigma_{wY}^2 + \sigma_{bY}^2)]}$ and variance $\sigma_{\mu_{X_i}}^2 = V(\mu_{X_i}) = \mu_X^2(e^{\sigma_{bY}^2} - 1)$. Figure 5.3B depicts the lognormal distribution of the random variable μ_{X_i}.

Here, $\sigma^2_{\mu_{X_i}}$ is the parameter reflecting between-person variability in the unobservable $\{\mu_{X_i}\}$. Thus, under Model (5.1), an observational group is also a *monomorphic group* [a group described by a lognormal distribution for μ_{X_i} (Rappaport, 1991b; Rappaport *et al.*, 1995a)], which can be used to make valid statistical inferences about the unobservable individual mean exposures, i.e., the $\{\mu_{X_i}\}$.

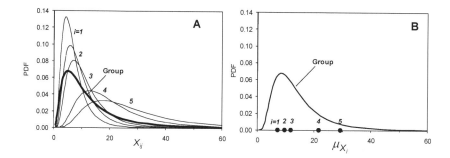

Fig. 5.3 Hypothetical lognormal distributions of exposure levels (arbitrary units) for an observational group of persons. A) Lognormal distributions of exposure levels received from day to day (X_{ij}). PDFs 1 – 5 represent a random sample of 5 persons from the group, while the PDF labelled 'Group' represents the hypothetical population of daily exposures for the entire group. B) Points labelled 1 – 5 represent the individual mean exposure levels (μ_{X_i}) of the corresponding persons depicted in A. The PDF labelled 'Group' represents the lognormal distribution of individual mean exposures $\{\mu_{X_i}\}$ for the group.

5.2.3 Estimating parameters

Model (5.1) can be fit to exposure data based on measuring n_i *repeated* exposure levels for the ith person in a random sample of k persons. Let x_{ij} represent the jth measured exposure concentration for the ith person, and let $y_{ij} = \ln(x_{ij})$ denote the corresponding logged measurement. Standard ANOVA methods can be applied to estimate the parameters under Model (5.1) as shown below (Searle *et al.*, 1992):

$y_i = \dfrac{1}{n_i}\sum_{j=1}^{n_i} y_{ij}$ is the sample mean of the logged measurements for the ith person;

$\bar{y} = \dfrac{1}{N}\sum_{i=1}^{k}\sum_{j=1}^{n_i} y_{ij}$ is the overall sample mean based upon all $N = \sum_{i=1}^{k} n_i$ logged measurements;

$\hat{\sigma}^2_{wY} = MSW$ (mean-square within) is the ANOVA estimate of the within-person variance component σ^2_{wY};

$$\hat{\sigma}^2_{bY} = \frac{(k-1)(MSB - MSW)}{\left(N - \sum_{i=1}^{k} n_i^2 / N\right)}$$ is the ANOVA estimate of the between-person variance component σ^2_{bY}; here, $MSW = \dfrac{\sum_{i=1}^{k}\sum_{j=1}^{n_i}(y_{ij} - \bar{y}_i)^2}{(N-k)}$ and

$$MSB = \frac{\sum_{i=1}^{k} n_i (\bar{y}_i - \bar{y})^2}{(k-1)}$$ (mean-square between) are obtained from the appropriate ANOVA table (discussed in Section 5.2.4); and

$$\hat{\mu}_Y = \frac{\sum_{i=1}^{k}\left(\bar{y}_i / (\hat{\sigma}^2_{bY} + \hat{\sigma}^2_{wY}/n_i)\right)}{\sum_{i=1}^{k}\left(1/(\hat{\sigma}^2_{bY} + \hat{\sigma}^2_{wY}/n_i)\right)}$$ estimates the true group mean μ_Y for all logged

measurements. Note that $\hat{\mu}_Y$ depends upon the estimated variance components and weights each $\bar{y}_i$ by the inverse of its estimated variance (namely, $\hat{\sigma}^2_{bY} + \frac{\hat{\sigma}^2_{wY}}{n_i}$), so that $\bar{y}_i$'s with smaller estimated variances are given more weight; and, $\hat{\sigma}^2_Y = \hat{\sigma}^2_{bY} + \hat{\sigma}^2_{wY}$ is the ANOVA estimate of the true variance σ^2_Y for all logged measurements. The above estimated parameters can be substituted in the appropriate conversion formulae for estimating the parameters of corresponding lognormal distributions. For example, the estimated overall mean and variance of the lognormal distribution of exposure measurements can be obtained as $\hat{\mu}_X = e^{[\hat{\mu}_Y + 0.5(\hat{\sigma}^2_{bY} + \hat{\sigma}^2_{wY})]}$ and $\hat{\sigma}^2_X = \hat{\mu}_X^2 (e^{(\hat{\sigma}^2_{bY} + \hat{\sigma}^2_{wY})} - 1)$, respectively.

Because MSW can be greater than MSB in some instances, the ANOVA estimate of σ^2_{bY}, i.e., $\hat{\sigma}^2_{bY} = \dfrac{(k-1)(MSB - MSW)}{\left(N - \sum_{i=1}^{k} n_i^2 / N\right)}$, is occasionally negative, particularly when k and $N = \sum_{i=1}^{k} n_i$ are small (Searle et al., 1992). In the event that $\hat{\sigma}^2_{bY} < 0$, $\hat{\sigma}^2_{bY}$ should then be set to zero, so that $\hat{\sigma}^2_Y = 0 + \hat{\sigma}^2_{wY} = \hat{\sigma}^2_{wY}$.

For the special case where only one exposure measurement is available for each person ($n_i = 1$, $i = 1, 2, \ldots, k$), then $N = \sum_{i=1}^{k} n_i = k$, and the above estimates of μ_Y and σ_Y^2 simplify to the familiar forms, $\hat{\mu}_Y = \frac{1}{k}\sum_{i=1}^{k} y_i$ and $\hat{\sigma}_Y^2 = \frac{1}{(k-1)}\sum_{i=1}^{k}(y_i - \bar{y})^2$, that were obtained for Model (4.1), where y_i denotes the single exposure measurement for subject i.

5.2.4 The ANOVA table

Following application of Model (5.1) with any standard statistical package, an ANOVA table is obtained that includes the following information:

Source	Sum of Squares	Degrees of Freedom (d.f.)	Mean Squares	Parameter Estimated
Between-person	SSB	$k-1$	$\dfrac{SSB}{(k-1)}$	$\sigma_{wY}^2 + n_0 \sigma_{bY}^2$
Within-person (error)	SSW	$N-k$	$\dfrac{SSW}{(N-k)}$	σ_{wY}^2

Here $SSB = \sum_{i=1}^{k} n_i (\bar{y}_i - \bar{y})^2$ and $SSW = \sum_{i=1}^{k}\sum_{j=1}^{n_i}(y_{ij} - \bar{y}_i)^2$ are the sums of squares between-persons and within-persons, respectively, and $n_0 = \dfrac{N - \dfrac{\sum_{i=1}^{k} n_i^2}{N}}{(k-1)}$ is the weighted average number of measurements per person (when $n_i = n$, $i = 1, 2, \ldots, k$, then $n_0 = n$ as expected). Assuming Model (5.1) holds, Searle et al. (1992) provide expressions for the estimated variances of σ_{bY}^2 and σ_{wY}^2, and for the estimated covariance between σ_{bY}^2 and σ_{wY}^2, so that confidence intervals and tests of hypotheses about important parameters can be constructed.

5.3 Relative measures of variability

The ranges of exposure levels indicated by $\hat{\sigma}_{bY}^2$ and $\hat{\sigma}_{wY}^2$ are often difficult to gauge because both $\hat{\sigma}_{bY}^2$ and $\hat{\sigma}_{wY}^2$ are computed using the log-transformed exposure measurements. To make exposure variability more interpretable, Rappaport (1991b) defined a scale-independent measure of exposure variability, $R_{0.95}$, representing the fold-range containing the middle 95% of the exposure concentration values. That is,

$$R_{0.95} = \frac{\text{97.5-th Percentile exposure level}}{\text{2.5-th Percentile exposure level}}.$$

This fold-range can be defined for either the underlying lognormal distribution of unobservable subject-specific means $\{\mu_{X_i}\}$ or the conditional lognormal distribution of daily exposures $\{X_{ij}\}$ given subject i (or, equivalently, given μ_{X_i}). For the distribution of μ_{X_i} (between-subject variability), since $\ln(\mu_{X_i})$ is $N(\mu_Y + \frac{\sigma_{wY}^2}{2}, \sigma_{bY}^2)$, the symmetric interval $(\mu_Y + \frac{\sigma_{wY}^2}{2} - 1.96\sigma_{bY}, \mu_Y + \frac{\sigma_{wY}^2}{2} + 1.96\sigma_{bY})$ contains 95% of the possible $\{\ln(\mu_{X_i})\}$ values. Hence, it is natural to define the fold-range containing 95% of the unlogged $\{\mu_{X_i}\}$ as $_b R_{0.95} = \frac{\mu_{X_i, 0.975}}{\mu_{X_i, 0.025}}$, where $\mu_{X_i, 0.975} = e^{[(\mu_Y + 0.5\sigma_{wY}^2) + 1.96\sigma_{bY}]}$ and $\mu_{X_i, 0.025} = e^{[(\mu_Y + 0.5\sigma_{wY}^2) - 1.96\sigma_{bY}]}$, so that $_b R_{0.95} = e^{3.92\sigma_{bY}}$. For example, if $_b R_{0.95} = e^{3.92\sigma_{bY}} = 4$, then 95% of the values for the lognormal distribution of μ_{X_i} would lie within a 4-fold range; this occurs when $\sigma_{bY} = 0.354$.

Likewise, for variation within persons, $Y_{ij} = \ln(X_{ij} | \mu_{X_i}) \sim N(\mu_{Y_i}, \sigma_{wY}^2)$, where $\mu_{Y_i} = (\mu_Y + b_i)$, so that the interval $(\mu_{Y_i} - 1.96\sigma_{wY}, \mu_{Y_i} + 1.96\sigma_{wY})$ contains 95% of the possible (logged) exposure values for subject i. Thus, we define $X_{0.975} | \mu_{X_i} = e^{(\mu_{Y_i} + 1.96\sigma_{wY})}$ and $X_{0.025} | \mu_{X_i} = e^{(\mu_{Y_i} - 1.96\sigma_{wY})}$, so that $_w R_{0.95} = e^{3.92\sigma_{wY}}$. For example, when $_w R_{0.95} = 15$, then 95% of the daily exposures experienced by any individual would have a 15-fold range; this occurs when $\sigma_{wY} = 0.691$.

For particular data sets, estimates of these fold-ranges can be obtained by substituting values of $\hat{\sigma}_{bY}$ and $\hat{\sigma}_{wY}$ into the above relationships. These estimates are designated $_b \hat{R}_{0.95} = e^{3.92\hat{\sigma}_{bY}}$ and $_w \hat{R}_{0.95} = e^{3.92\hat{\sigma}_{wY}}$, respectively. Table 5.2 lists values of $_b \hat{R}_{0.95}$ and $_w \hat{R}_{0.95}$, along with the corresponding variance components estimated for Groups 1 – 4. When considering exposure variability both between and within persons, we designate the overall fold-range as $_Y R_{0.95} = e^{3.92\sqrt{\sigma_{bY}^2 + \sigma_{wY}^2}}$ and its estimate as $_Y \hat{R}_{0.95} = e^{3.92\sqrt{\hat{\sigma}_{bY}^2 + \hat{\sigma}_{wY}^2}}$.

Table 5.2 Parameter estimates for Groups 1 – 4 [ANOVA estimates from application of Model (5.1) to the data].

Group	$\hat{\mu}_Y$	$\hat{\sigma}^2_{bY}$ ($_b\hat{R}_{0.95}$)	$\hat{\sigma}^2_{wY}$ ($_w\hat{R}_{0.95}$)	$\hat{\sigma}^2_{Y}$ ($_Y\hat{R}_{0.95}$)	$\hat{\mu}_X$	$\hat{\rho} = \dfrac{\hat{\sigma}^2_{bY}}{\hat{\sigma}^2_Y}$
1	2.73	0.023 (1.82)	0.377 (11.1)	0.400 (12.0)	18.7 µg/m³	0.058
2	-0.423	0.033 (2.03)	0.466 (14.5)	0.499 (16.0)	0.841 mg/m³	0.066
3	-6.85×10^{-2}	2.32 (392)	2.89 (782)	5.21 (7673)	12.6 mg/m³	0.445
4	4.62	0.292 (8.31)	0.513 (16.6)	0.805 (33.7)	152 mg/m³	0.362

5.4 Ranges of variance components in occupational groups

Table 5.2 lists the parameter estimates for Groups 1 – 4 following application of Model (5.1) to the data. From the table, we see that both $\hat{\sigma}^2_{bY}$ and $\hat{\sigma}^2_{wY}$ covered wide ranges; in fact, extremely large estimates of both variance components were observed for Group 3. The last column shows the estimated intraclass correlation, represented by the ratio of the estimated between-person variance component to the estimated overall variance $\hat{\rho} = \hat{\sigma}^2_{bY} / \hat{\sigma}^2_Y$; these values ranged from 0.058 for Group 1 to 0.445 for Group 3 with an average of 0.233. The results from Groups 1-4 are consistent with those for a large database of occupational exposures from 220 observational groups, where the median value of $\hat{\rho}$ was 0.22, 25% of the groups had $\hat{\rho} \leq 0.04$, and 25% had $\hat{\rho} \geq 0.41$ (Tornero-Velez et al., 1997). Since the estimated intraclass correlation is greater than zero in most data sets, it appears that repeated measurements from the same person tend to be positively correlated. As mentioned previously, statistical methods that do not take into account such correlation can lead to invalid statistical conclusions.

In an extensive investigation of variability of occupational exposures, based upon 13,945 daily exposure concentrations (personal measurements) for 1,574 workers, Kromhout et al. (1993) reported values of $\hat{\sigma}^2_{wY}$ and $\hat{\sigma}^2_{bY}$ for 165 groups of workers classified by job and location (factory). The cumulative distributions of these estimated variance components (reported as values of $_b\hat{R}_{0.95}$ and $_w\hat{R}_{0.95}$) are shown in Figure 5.4. The figure shows that exposure levels tend to vary more within persons than between persons ($_w\hat{R}_{0.95}$ median = 15.1, $_b\hat{R}_{0.95}$ median = 4.06), a finding consistent with more recent studies (Kumagai et al., 1996; Peretz et al., 1997; Woskie et al., 1994). However, estimates of $_wR_{0.95}$ and $_bR_{0.95}$ both covered extremely wide ranges (from less than 2 to greater than 1000), suggesting many sources of variability operating at both between-person and within-person levels.

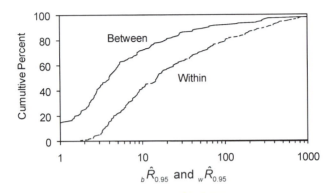

Fig. 5.4 Cumulative distributions of within-person and between-person sources of variation in exposure for 165 observational groups of workers [from Kromhout *et al.* (1993)]. (Expressed as the estimated fold-ranges $_b\hat{R}_{0.95}$ and $_w\hat{R}_{0.95}$).

5.4.1 Sources of variability in occupational groups

As part of their study, Kromhout et al. (1993) evaluated the influence of several environmental and process-related covariates upon $\hat{\sigma}^2_{wY}$ and $\hat{\sigma}^2_{bY}$ (in the 87 observational groups with complete information). Univariate analyses showed that the following covariates were significantly associated with increasing exposure variability: intermittent processes > continuous processes (*P*-value < 0.001); outdoor exposures > indoor exposures (*P*-value < 0.01); general ventilation > local-exhaust ventilation (*P*-value < 0.01); mobile workers > stationary workers (*P*-value < 0.05); and local sources > general sources (*P*-value < 0.05). These associations were all in the expected directions.

Regression models were then used to investigate the combined effects of these environmental and process-related factors upon $\hat{\sigma}^2_{wY}$ and $\hat{\sigma}^2_{bY}$ (Kromhout *et al.*, 1993). The results, summarized in Table 5.3, showed that the final model for $\hat{\sigma}^2_{wY}$ contained only two effects (continuous vs. intermittent process and indoor vs. outdoor exposure), while that for $\hat{\sigma}^2_{bY}$ contained only a single effect (continuous vs. intermittent process). For each of the variance components given in the table, the model-based predicted values are shown for the indicated combination of covariates. Note that only three combinations of covariates were available for predictions because no continuous outdoor processes were represented in the 87 occupational groups (with complete information) used to develop the regression models.

These results suggest that variability within persons is largely dictated by factors related to the process and the environment, a finding confirmed by Peretz et al. (1997). Since such factors would affect all persons in the group more or less equally, the fact that they explain 41% of the variability in $\hat{\sigma}_{wY}^2$ (Table 5.3) is, perhaps, not surprising. On the other hand, these same factors explained only 13% of the variation in $\hat{\sigma}_{bY}^2$ and the fit of the regression model was rather poor.

Table 5.3 Regression evaluation of covariates on estimated within-person and between-person variance components for 87 occupational groups [from Kromhout et al. (1993)].

Dependent Variable	Covariate(s)	R^2 (%)	Model Prediction of Var. Comp.	Model Prediction of Fold-Range
$\hat{\sigma}_{wY}^2$	Process and environment	41	σ_{wY}^2	$_wR_{0.95}$
	Continuous and indoors		0.320	9.2
	Intermittent and indoors		1.30	87.6
	Intermittent and outdoors		1.60	142
$\hat{\sigma}_{bY}^2$	Process	13	σ_{bY}^2	$_bR_{0.95}$
	Continuous		0.053	2.5
	Intermittent		0.320	9.2

In a more recent study of within-worker and between-worker variability in the rubber industry, Peretz et al. (2002) identified 31 tasks and processes that significantly explained exposure variability. When all these covariates were included in the model of exposure in the rubber industry, the estimate of $\hat{\sigma}_{bY}^2$ was reduced by 35%, whereas that for $\hat{\sigma}_{wY}^2$ was unchanged. This indicates that between-person variability in workplace exposure levels can largely be explained by the differences in tasks and locations experienced by individuals in a given job. Other variables, such as types of equipment used by workers in a given job and particular work practices, would also be expected to explain variability in $\hat{\sigma}_{bY}^2$.

The predicted variance components in Table 5.3 strongly suggest that intermittent processes are typically much more variable than continuous processes, both within and between workers. As we shall see in Chapter 9, this finding has implications for sample size requirements for testing exposure levels relative to OELs.

5.4.2 Uniformity of mean exposure levels across persons

The wide range of $\hat{\sigma}_{bY}^2$ values reported in the above studies casts further doubt upon the traditional model of exposure, depicted by Model (4.1). Under Model (4.1), subject-specific mean exposures do not vary across persons; i.e., it is assumed that the intraclass correlation of the group $\rho = \sigma_{bY}^2 / \sigma_Y^2 = 0$. Based upon our current knowledge, this is unreasonable because evidence indicates that $\rho > 0$ in the majority of observational groups.

Since uniform exposure occurs when the variability between persons is very small (i.e., $\hat{\sigma}_{bY}^2$ approaches 0 or $_b\hat{R}_{0.95}$ approaches one), we can arbitrarily regard a group to be uniformly exposed when $_b\hat{R}_{0.95} \leq 2$ (or equivalently when $\hat{\sigma}_{bY}^2 \leq 0.0313$) (Rappaport, 1991b). Referring now to the estimated between-person variability for Groups 1 – 4 (Table 5.2), Groups 1 and 2 provide evidence of uniform mean exposures across persons since $_b\hat{R}_{0.95}$ = 1.8 and 2.0, respectively. However, Groups 3 and 4, with $_b\hat{R}_{0.95}$ = 421 and 8.3, respectively, provide evidence of non-uniform mean exposures across persons.

Uniformity of mean exposure levels across persons can be considered more generally based on the compilation of between-person variance components reported by Kromhout *et al.* (1993) for 165 observational groups (see Figure 5.4). Based on these estimates, values of $_b\hat{R}_{0.95}$ ranged from 1 to 2000 with a median value of 4.06. This median value indicates that workers in a typical observational group experienced about a 4-fold range of individual mean exposures. Only about 20% of these groups had values of $_b\hat{R}_{0.95} < 2$, suggesting uniform mean exposure across subjects; an equal percentage had values of $_b\hat{R}_{0.95} > 18$, suggesting fairly large heterogeneity in mean exposure levels across workers.

With strong empirical evidence that about 4/5 of observational groups are non-uniformly exposed, we should be wary of using a statistical model for exposure data that cannot accurately quantify the within-person and between-person variability in exposure levels. Indeed, the use of Model (4.1) is inappropriate for the valid assessment of sources of variability for nearly all datasets of occupational and environmental exposures. Its use can actually hamper any systematic investigations of differences in exposure levels due to tasks, practices, and personal environments.

5.5 Environmental exposures

Although Model (5.1) has been applied extensively in occupational studies, it has rarely been used to describe exposure distributions in environmental studies. Rappaport and Kupper (2004) applied this model to air concentrations of volatile organic compounds (VOCs) reported by the U.S. Environmental Protection Agency in the TEAM studies of the 1980s, where 12-h day and night personal measurements were obtained from a stratified sample of hundreds of residents from 5 communities in the U. S. Since some of these subjects were monitored on multiple days, it was possible to estimate σ_{wY}^2 and σ_{bY}^2 from these data and then compare these estimates with the corresponding estimated variance components reported by Kromhout *et al.* (1993) for the 12 VOCs from their database of occupational studies.

These comparisons are illustrated in Figure 5.5, which shows cumulative distributions of $_w\hat{R}_{0.95}$ and $_b\hat{R}_{0.95}$ corresponding to $\hat{\sigma}_{bY}^2$ (left) and $\hat{\sigma}_{wY}^2$ (right), respectively. Regarding the cumulative distributions of $_b\hat{R}_{0.95}$, fold ranges were similar for environmental and occupational exposures (medians ranged roughly between 3.3 and 4.0), although daytime exposures were noticeably more variable than both nighttime exposures and occupational exposures in about half of the datasets. For the cumulative distributions of $_w\hat{R}_{0.95}$, fold ranges were much greater for environmental exposures (day median = 136, night median = 80) than for occupational exposures (median = 15), and day exposures were more variable within persons than were night exposures in about 80% of the cases. Rappaport and Kupper (2004) concluded that the variation in mean exposure levels for VOCs across residents in a given city was comparable to that observed across workers in a given factory and job. However, the greatly increased within-person variation in environmental exposures suggests that sample sizes in community studies should almost certainly be larger than those in occupational studies in order to validly and precisely quantify sources of exposure variability and health-related risks. This subject is dealt with extensively in Chapter 10.

In another application of random effects models to environmental exposures, Symanski *et. al.* (2004) reported estimated variance components for exposures to disinfection byproducts in water from households, measured during different seasons of the year. In that investigation, inter-household variation (analogous to $\hat{\sigma}_{bY}^2$) in levels of trihalomethanes (one class of disinfection byproducts) represented more than half of the total variability, after adjusting for the effect of season of the year. This indicates that factors affect the transport of disinfection byproducts to different households from a common municipal water supply.

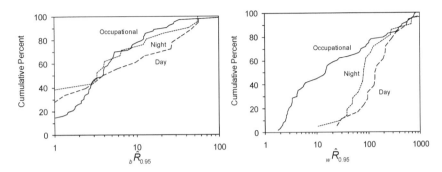

Fig. 5.5 Cumulative distributions of fold-ranges of personal VOC exposure levels between-subjects and within-subjects estimated under Model (5.1). 'Day' and 'Night' refer to 12-h environmental exposures of 9 VOCs from measurements reported by the TEAM study during daytime and nighttime, respectively (Rappaport and Kupper, 2004). 'Occupational' refers to shift-long personal measurements to 12 VOCs reported by Kromhout *et al.* (1993). Left: Estimated between-person fold-range; Right: Estimated within-person fold-range.

5.6 This chapter and Chapter 6

In this chapter, we introduced the simplest random-effects model [Model (5.1)] and used it to estimate within-person and between-person components of variance in exposure levels for various groups of persons. Using databases of contaminant levels in the workplace and the general environment, we showed that both of these variance components can be very large in observational groups. In Chapter 6, we will consider more general statistical models (linear mixed effects models) that can be used to simultaneously estimate within-person and between-person variance components across multiple groups and can also help to evaluate the effects of important determinants of exposure levels.

6 MIXED MODELS OF EXPOSURE

6.1 General linear mixed models

In Chapter 5, we discussed how the one way random effects model [Model (5.1)] has been used to describe exposure variability within and between persons in a single group. Recalling the form of Model (5.1), $Y_{ij} = \ln(X_{ij}) = \mu_Y + b_i + e_{ij}$, we see that it is a *linear mixed model*, where μ_Y is the fixed true mean of the logged exposure levels for the group [i.e., $\mu_Y = E(Y_{ij})$], and where b_i and e_{ij} represent independent random effects for the ith person and the jth measurement on the ith person, respectively. Thus, Model (5.1) is one member of the family of general linear mixed models that contain both fixed and random effects.

Following the development of computationally efficient and readily available software packages, general linear mixed models are finding increasing use in analyses of correlated data, such as data involving repeated exposure measurements on each of several persons. Given the importance of these models for comprehensive analyses of exposure data, let us digress a moment and consider notational conventions for defining such models.

6.1.1 Matrix notation

The general linear mixed model is specified as:

$$Y = X\beta + Zb + e. \tag{6.1}$$

Here, **Y** is the vector of response random variables (e.g., logged exposure levels), **X** is the known design matrix for fixed effects, **β** is the matrix of unknown fixed effects parameters, **Z** is the known design matrix for random effects, **b** is the vector of unobservable random effects, and **e** is the vector of unobservable random errors. It is assumed that **b** and **e** are independent of each other and are multivariate normal with expected value vector $E\begin{bmatrix} \mathbf{b} \\ \mathbf{e} \end{bmatrix} = \begin{bmatrix} \mathbf{0} \\ \mathbf{0} \end{bmatrix}$ and variance-covariance matrix $V\begin{bmatrix} \mathbf{b} \\ \mathbf{e} \end{bmatrix} = \begin{bmatrix} \mathbf{G} & \mathbf{0} \\ \mathbf{0} & \mathbf{R} \end{bmatrix}$. Thus, $V(\mathbf{b}) = \mathbf{G}$, $V(\mathbf{e}) = \mathbf{R}$, and so $V(\mathbf{Y}) = \mathbf{ZGZ'} + \mathbf{R}$. Hence, the fixed effects are modeled with **X** and **β**, the random effects with **Z**, **b**, and **G**, and the variance-covariance matrix of **e** with **R**.

To illustrate how the matrices defined in Model (6.1) are constructed, consider a simple example involving some data just from Group 2, which was described in Chapter 5. Suppose that only the first two measured exposure concentrations ($n = 2$) for each of the first three subjects ($k = 3$) from Group 2 were available for analysis. These observed data $\{x_{ij}\}$, and their corresponding natural logarithms $\{y_{ij}\}$, are shown in Table 6.1:

Table 6.1 Example data to illustrate mixed Model (6.1). (These data are comprised of the first two exposure measurements from each of the first three persons in Group 2).

i (Person)	j (Day)	x_{ij} (mg/m³)	y_{ij}
1	1	0.48	-0.734
1	2	0.32	-1.139
2	1	0.54	-0.616
2	2	1.09	0.086
3	1	0.42	-0.868
3	2	0.28	-1.273

Under the general structure of Model (6.1), a one way random effects model is depicted for the 6 random variables Y_{11}, Y_{12}, Y_{21}, Y_{22}, Y_{31}, and Y_{32}, as follows:

$$\mathbf{Y} = (\mathbf{X}\,\boldsymbol{\beta}) + (\mathbf{Z}\,\mathbf{b}) + \mathbf{e}$$

$$\begin{bmatrix} Y_{11} \\ Y_{12} \\ Y_{21} \\ Y_{22} \\ Y_{31} \\ Y_{32} \end{bmatrix} = \begin{bmatrix} 1 \\ 1 \\ 1 \\ 1 \\ 1 \\ 1 \end{bmatrix} [\mu_Y] + \begin{bmatrix} 1 & 0 & 0 \\ 1 & 0 & 0 \\ 0 & 1 & 0 \\ 0 & 1 & 0 \\ 0 & 0 & 1 \\ 0 & 0 & 1 \end{bmatrix} \begin{bmatrix} b_1 \\ b_2 \\ b_3 \end{bmatrix} + \begin{bmatrix} e_{11} \\ e_{12} \\ e_{21} \\ e_{22} \\ e_{31} \\ e_{32} \end{bmatrix}.$$

Under Model (5.1), recall that b_i and e_{ij} are assumed to be independent random variables for all i and j, with zero means and with variances σ_{bY}^2 and σ_{wY}^2, respectively, so that $\sigma_Y^2 = \sigma_{bY}^2 + \sigma_{wY}^2$. It then follows that $\text{Cov}(Y_{ij}, Y_{ij'}) = \sigma_{bY}^2$ for all data pairs from the same person, where $j \neq j'$. These variances and covariances define the matrices **G** and **R** as follows:

$$V(\mathbf{b}) = \mathbf{G} = \sigma_{bY}^2 \mathbf{I}_3, \text{ where } \mathbf{I}_3 \text{ is the (3x3) identity matrix;}$$
$$V(\mathbf{e}) = \mathbf{R} = \sigma_{wY}^2 \mathbf{I}_6 \text{ ; and,}$$

$$V(Y) = V = ZGZ' + R = \sigma_{bY}^2 \begin{bmatrix} J_2 & 0_2 & 0_2 \\ 0_2 & J_2 & 0_2 \\ 0_2 & 0_2 & J_2 \end{bmatrix} + \sigma_{wY}^2 I_6,$$

where J_2 is a (2x2) matrix containing all ones, and 0_2 is a (2x2) matrix containing all zeros. For the subset of data from Group 2, the estimated variance-covariance matrix for the ith subject is $\hat{V}_i = \hat{\sigma}_{bY}^2 J_2 + \hat{\sigma}_{wY}^2 I_2 = \begin{bmatrix} 0.2547 & 0.1177 \\ 0.1177 & 0.2547 \end{bmatrix}$, where $\hat{\sigma}_{bY}^2 = 0.1177$, $\hat{\sigma}_{wY}^2 = 0.1370$, and $\hat{\sigma}_Y^2 = 0.2547$. More generally, for the one way random effects model with n measurements for each of k randomly selected subjects, $\hat{V}_i = \hat{\sigma}_{bY}^2 J_n + \hat{\sigma}_{wY}^2 I_n$, $i = 1, 2, ..., k$. This type of variance-covariance structure is referred to as *compound symmetry*. It is a common structure employed when fitting linear mixed models, and it is the one that we will generally use in our analyses. However, other structures are possible and may be desired in some circumstances [Searle *et al*. (1992) provide details].

6.2 Fitting linear mixed models separately to Groups 1 – 4

6.2.1 Profile plots

A useful preliminary step in applying linear mixed models to describe exposure data is to sequentially plot the values of repeated (logged) exposure measurements for each subject in a random sample of persons from a particular group. Such *profile plots* provide visual information with which to gauge possible trends in exposure levels over time, to view within-person and between-person variability in the data, and to identify possible outliers. To illustrate, Figure 6.1 shows profile plots for the data from Groups 1 – 4, where values of $y_{ij} = \ln(x_{ij})$ are plotted versus the serial order of measurements for each subject, which we will refer to as the *repeat* variable.[12] For Group 1 (inorganic lead) and Group 3 (benzene), the plots are unremarkable, with no discernable time trends in logged exposure levels during the periods of sampling. However, for Group 2 (benzene), the plot possibly suggests a declining trend in logged exposure levels, while that for Group 4 (styrene) suggests a slight upward trend in logged exposure levels over time.

---

[12] For the *repeat* variable to be most informative about time trends in exposure levels, data for all subjects (in a given group) should be collected at the same equally-spaced times, and there should be no missing data. Since these conditions were not fully satisfied for the data from these 4 groups, the *repeat* variable is used here primarily for illustration purposes and should be interpreted cautiously as a surrogate for time.

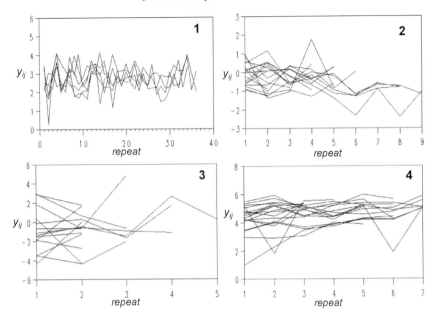

Fig. 6.1 Profile plots of logged exposure measurements (y_{ij}) shown by subject for Groups 1 – 4 (indicated by numbers). Plots show values of logged exposure levels versus the serial order of collection (*repeat*).

6.2.2 REML estimates

Since closed-form expressions for parameter estimates are only derivable for the simplest linear mixed models (with balanced data), most applications require the use of sophisticated statistical software, such as Proc MIXED of SAS (which employs iterative procedures to maximize the likelihood function). Of the several types of estimation methods available, restricted maximum likelihood (REML) estimation is regarded as the most appropriate for our applications (Searle *et al.*, 1992) and will be used here. In Chapter 5, we obtained ANOVA estimates for key parameters, using the data for Groups 1 - 4 (Table 5.1). Employing Proc MIXED of SAS to fit Model (5.1) *separately* to the data for each group, the corresponding REML estimates are shown in Table 6.2; these estimates are quite similar to the ANOVA estimates obtained previously for these four data sets. [For balanced data, REML and ANOVA estimates would be the same under Model (5.1)].

Table 6.2 REML and ANOVA parameter estimates for Groups 1 – 4.

Group	$\hat{\mu}_Y$		$\hat{\sigma}^2_{bY}$		$\hat{\sigma}^2_{wY}$	
	REML	ANOVA	REML	ANOVA	REML	ANOVA
1	2.73	2.73	0.023	0.023	0.377	0.377
2	-0.423	-0.423	0.040	0.033	0.459	0.466
3	-6.89×10^{-2}	-6.85×10^{-2}	2.45	2.32	3.07	2.89
4	4.62	4.62	0.283	0.292	0.509	0.513

6.2.3 Normality of predicted random effects

Estimated random effects (i.e., estimates of the $\{b_i\}$, designated $\{\hat{b}_i\}$) and their estimated standard errors can be generated for each subject in the random sample. The $\{\hat{b}_i\}$ generated by Proc MIXED are referred to as *empirical Bayes linear unbiased predictors* (EBLUPs). These estimators $\{\hat{b}_i\}$ have particular statistical properties that make them useful for evaluating the validity of assumptions (e.g., normality) regarding the random effects in a linear mixed model (Searle *et al.*, 1992).

One method for assessing the validity of the assumption of normality for the random-person effects (i.e., the b_i's) involves visual inspection of the *standardized estimators*, i.e., the $\{\hat{b}_i\}$ divided by their standard errors, namely, the $\{\hat{b}_i/\text{SE}(\hat{b}_i)\}$. When the $\{\hat{b}_i/\text{SE}(\hat{b}_i)\}$ are plotted in either a normal probability or q-q (quantile-quantile) format, the reasonableness of the normality assumption is supported by an approximate straight-line relationship. One can also apply a test for normality, such as the Shapiro-Wilks W-test; however, the *P*-values for such tests should be regarded cautiously because the $\{\hat{b}_i/\text{SE}(\hat{b}_i)\}$ are not adjusted for estimation of the unknown parameters upon which their computation is based (Lange and Ryan, 1989). Figure 6.2 shows plots of the standardized estimators $\{\hat{b}_i/\text{SE}(\hat{b}_i)\}$ obtained from Proc MIXED for Groups 1 – 4 under the one way random effects model [Model (5.1)], and Table 6.3 provides results from the corresponding Shapiro-Wilks W-tests. In all cases, the plots suggest that the assumption of normality of the $\{b_i\}$ is not unreasonable, especially for Groups 1 – 3. The large *P*-values of the W-tests for Groups 1 – 3 (0.346 – 0.829) support these graphical results. For Group 4, the two smallest values of $\hat{b}_i/\text{SE}(\hat{b}_i)$, corresponding to subjects 1 and 7, deviate marginally from the straight-line relationship, and the *P*-value of the W-test roughly approaches statistical significance (*P*-value = 0.124).

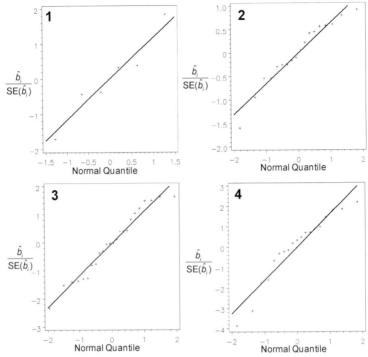

Fig. 6.2 Quantile-quantile plots of standardized subject-specific random effects from Groups 1 – 4 (indicated by numbers). In each plot, $\hat{b}_i/\text{SE}(\hat{b}_i)$ is the estimated random effect $\hat{b}_i$ for the ith subject divided by its estimated standard error, $\text{SE}(\hat{b}_i)$.

Table 6.3 Assessment of the assumption of normality of $\{b_i\}$ under Model (5.1) for Groups 1-4 using the Shapiro-Wilks W statistic to test the $\{\hat{b}_i/\text{SE}(\hat{b}_i)\}$.

Group	No. Persons (k)	W Statistic	P-value
1	6	0.961	0.829
2	18	0.948	0.388
3	24	0.955	0.346
4	19	0.922	0.124

6.2.4 Normality of residuals

For the class of general linear mixed models defined by Model (6.1), the vector of predicted responses is defined as $\hat{\mathbf{y}} = \mathbf{X}\hat{\boldsymbol{\beta}} + \mathbf{Z}\hat{\mathbf{b}}$, and $\hat{\mathbf{e}} = (\mathbf{y} - \hat{\mathbf{y}})$ is the set of conditional residuals. More specifically, for the one way random effects model [Model (5.1)], a residual specific to Subject i is defined as

$\hat{e}_{ij} = (y_{ij} - \hat{y}_{ij} \mid \text{Subject } i) = (y_{ij} - \hat{\mu}_{Y_i})$, where $\hat{\mu}_{Y_i} = (\hat{\mu}_Y + \hat{b}_i)$. Since, under Model (5.1), the set of residuals $\{\hat{e}_{ij}\}$ should roughly behave like a random sample from a normal distribution, a q-q plot of the $\{\hat{e}_{ij}\}$ for a set of data should follow an approximate straight line relationship.

Figure 6.3 shows q-q plots of the residuals for Groups 1-4. Visually, the data from the first three groups do not appear to violate the normality assumption. However, one observation from Group 1 and three observations from Group 4 are much smaller than predicted under Model (5.1). Conclusions from the plotted data are reinforced by results (see Table 6.4) from application of the Shapiro Wilks W-test to the sets of $\{\hat{e}_{ij}\}$ for Groups 1 - 4; i.e., P-values for Groups 2 and 3 are 0.1955 and 0.5894, respectively, while those for Groups 1 and 4 are 0.0175 and <0.0001, respectively.

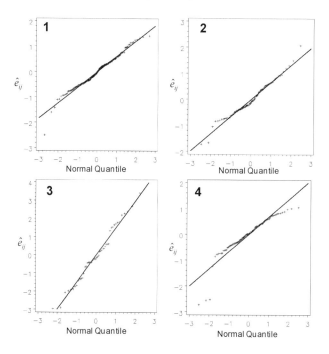

Fig. 6.3 Quantile-quantile plots of residuals under Model (5.1) for Groups 1 – 4. In each plot, the residual ($\hat{e}_{ij}$) represents the deviation of the jth observation from the ith person from that predicted under Model (5.1).

Table 6.4 Assessment of the assumption of normality of the $\{e_{ij}\}$ under Model (5.1) for Groups 1-4 using the Shapiro-Wilks W Statistic for the $\{\hat{e}_{ij}\}$.

Group	Number of Observations (N)	W Statistic	P-value
1	177	0.9813	0.0175
2	90	0.9805	0.1955
3	48	0.9803	0.5894
4	103	0.8768	<0.0001

6.3 Modeling data involving several groups and covariates

Since it can be more statistically efficient to simultaneously model exposure data from several groups and, at the same time, to evaluate the effects of covariates upon exposure levels, we will now consider more complex mixed models [designated Models (6.2) and (6.3)] that permit such analyses. Using Models (6.2) and (6.3), we will estimate fixed effects associated with different groups and with different covariates (e.g., job, time, and environment-specific or task-specific factors), while also estimating the within-person and between-person variance components. Prior to the use of linear mixed models, investigators relied upon a two-step process to model exposure data; in particular, variance components were estimated separately from fixed effects, using statistical models which were probably not optimal for either type of estimation process (Kromhout et al., 1994; Lagorio et al., 1998; Preller et al., 1995; Woskie et al., 1994).

Multiple groups can be investigated with the following linear mixed model:

$$Y_{hij} = \ln(X_{hij}) = \mu_{Y_h} + b_{hi} + e_{hij} \tag{6.2}$$

for $h = 1, 2, ..., H$ groups,
for $i = 1, 2, ..., k_h$ persons in Group h,
and for $j = 1, 2, ..., n_{hi}$ exposure measurements for person i in Group h.

Here, X_{hij} represents the exposure concentration on the jth day for person i in Group h, where j again indexes time (e.g., day j). The natural logarithm of X_{hij}, namely $Y_{hij} = \ln(X_{hij})$, is modeled as the sum of both fixed and random effects, defined as follows: μ_{Y_h} is the true underlying fixed (logged) mean exposure level for Group h; b_{hi} is the random effect of person i in Group h [i.e., $b_{hi} = (\mu_{Y_{hi}} - \mu_{Y_h})$, where $\mu_{Y_{hi}}$ is the true underlying mean of the (logged) exposure levels for person i in Group h]; and, e_{hij} is the random error for the jth day for person i [i.e., $e_{hij} = (Y_{hij} - \mu_{Y_{hi}})$]. Using the notation for the general linear mixed model [Model (6.1), i.e., **Y** = **Xβ**+**Zb**+**e**] for H groups, **Y** represents the vector of observed values for the $\sum_{h=1}^{H}\sum_{i=1}^{k_h} n_{hi}$ response random variables (Y_{111}, Y_{112}, ..., $Y_{Hk_H n_{Hk_H}}$), **β** represents the vector of unknown fixed group means (μ_{Y_1}, μ_{Y_2}, ..., μ_{Y_H}), **b** represents the vector of unknown random person effects (b_{11}, b_{12}, ... b_{Hk_H}), and **e** represents the vector of unknown random error effects (e_{111}, e_{112}, ... $e_{Hk_H n_{Hk_H}}$).

Model (6.3) differs only slightly from Model (6.2) by adding U covariates C_1, C_2, ..., C_U, as follows:

$$Y_{hij} = \ln(X_{hij}) = \mu_{Y_h} + \sum_{u=1}^{U} \delta_{uhj} C_{uhij} + b_{hi} + e_{hij} \quad (6.3)$$

for h = 1, 2, ..., H groups,
for i = 1, 2, ..., k_h persons in Group h,
for j = 1, 2, ..., n_{hi} measurements for person i in Group h,
and for u = 1, 2, ..., U covariates.

Here, the $\{\delta_{uhj}\}$ are regression coefficients representing the fixed effects of the U covariates. Under Model (6.3), note that the covariate effects $\{\delta_{uhj}\}$ are allowed to vary by group and by time, but not by subject; then, E(Y_{hij}) = $\mu_{Y_h} + \sum_{u=1}^{U} \delta_{uhj} C_{uhij}$. As important covariates are added to Model (6.3), the *explained* variation in logged exposure levels increases, thereby diminishing the variability attributable to the within-person and/or between-person components of variance.

Several applications of Models (6.2) and (6.3) suggest that these mixed models will be very useful in sorting out the many possible determinants of exposure and in selecting among options for interventions (Burstyn and Kromhout, 2000; Egeghy *et al.*, 2002; Egeghy *et al.*, 2000; Kromhout and Vermeulen, 2001; Mikkelsen *et al.*, 2002; Peretz *et al.*, 2002; Rappaport *et al.*, 1999; Symanski *et al.*, 2001; Symanski *et al.*, 2000; van Tongeren and Gardiner, 2001; Vermeulen *et al.*, 2001; Vermeulen *et al.*, 2000; Weaver *et al.*, 2001).

6.3.1 Assumptions

Under Models (6.2) and (6.3), it is assumed that the $\{b_{hi}\}$ and $\{e_{hij}\}$ are all mutually independent and normally distributed random variables with means of zero and variances of $\sigma_{bY_h}^2$ and $\sigma_{wY_h}^2$, representing the between-person and within-person components of variance, respectively, for Group h. It follows that Y_{hij} is normally distributed with mean μ_{Y_h} [Model (6.2)] or $\mu_{Y_h} + \sum_{u=1}^{U} \delta_{uhj} C_{uhij}$ [Model (6.3)] and variance $\sigma^2_{Y_h} = \sigma^2_{bY_h} + \sigma^2_{wY_h}$. The following statements summarize some assumptions about the means, variances, and covariances in Models (6.2) and (6.3) for subject i in Group h:

1. $E(Y_{hij}|\mu_{Y_{hi}}) = \mu_{Y_{hi}} = \mu_{Y_h} + b_{hi}$ [Model (6.2)] or

    $E(Y_{hij}|\mu_{Y_{hi}}) = \mu_{Y_{hi}} = \mu_{Y_h} + \sum_{u=1}^{U} \delta_{uhj} C_{uhij} + b_{hi}$ [Model (6.3)].

2. $V(Y_{hij}|\mu_{Y_{hi}}) = \sigma^2_{wY_h}$.

3. For $j \neq j'$, $Cov(Y_{hij}, Y_{hij'}) = \sigma^2_{bY_h}$, so that $Corr(Y_{hij}, Y_{hij'}) = \rho_h = \dfrac{\sigma^2_{bY_h}}{\sigma^2_{Y_h}}$.

It is also implicitly assumed under Model (6.2) that each exposure level (X_{hij}) for a person in Group h is lognormally distributed with mean $\mu_{X_h} = e^{(\mu_{Y_h} + 0.5\sigma^2_{Y_h})}$ and variance $\sigma^2_{X_h} = \mu^2_{X_h}(e^{\sigma^2_{Y_h}} - 1)$. Further, the person-specific mean exposures in Group h, i.e., the random variables $\mu_{X_{hi}} = e^{(\mu_{Y_{hi}} + 0.5\sigma^2_{wY_h})}$, are assumed to be lognormally distributed with mean μ_{X_h} and variance $\mu^2_{X_h}(e^{\sigma^2_{bY_h}} - 1)$. Analogous expressions are easily obtained under Model (6.3).

6.3.2 Estimating parameters

The estimates of the parameters μ_Y, μ_{Y_h}, $\sigma^2_{bY_h}$, and $\sigma^2_{wY_h}$ are designated $\hat{\mu}_Y$, $\hat{\mu}_{Y_h}$, $\hat{\sigma}^2_{bY_h}$, and $\hat{\sigma}^2_{wY_h}$, respectively. These estimated parameters are used to obtain $\hat{\sigma}^2_{Y_h} = \hat{\sigma}^2_{bY_h} + \hat{\sigma}^2_{wY_h}$, $\hat{\mu}_{X_h} = e^{(\hat{\mu}_{Y_h} + 0.5\hat{\sigma}^2_{Y_h})}$, and $\hat{\sigma}^2_{X_h} = \hat{\mu}^2_{X_h}(e^{\hat{\sigma}^2_{Y_h}} - 1)$. Again, REML estimates are generally preferred.

6.3.3 Estimating variance components across groups

When estimating variance components for several observational groups, it can be beneficial to assume common between-person and/or within-person variance components, when justified by the data (Rappaport et al., 1999; Weaver et al., 2001). Assuming common variance components across groups permits the variance components to be estimated even when individual groups have small numbers of observations. The following three alternative variance structures can be evaluated under Model (6.2):

1. Model (6.2A): $\sigma_{bY_h}^2$ and $\sigma_{wY_h}^2$ are assumed to vary across groups;
2. Model (6.2B): $\sigma_{bY_h}^2$ is assumed to vary across groups and $\sigma_{wY_h}^2 = \sigma_{wY}^2$ is assumed not to vary across groups; and,
3. Model (6.2C): $\sigma_{bY_h}^2 = \sigma_{bY}^2$ and $\sigma_{wY_h}^2 = \sigma_{wY}^2$ are both assumed not to vary across groups.

Model (6.2A) can be called the *full model* (or alternatively the unrestricted or unreduced model) since it allows both the within-person and the between-person variance components to vary from group to group. Variance components estimated via Model (6.2A) would be the same as those obtained by applying Model (5.1) to each group separately. Models (6.2B) and (6.2C) add *restrictions* in the form of assumptions of common within-person and/or between-person variance components, thereby reducing the total number of model parameters (and are, therefore, referred to as *reduced models*). That is, the number of variance parameters would be 2H (where H designates the number of groups) for Model (6.2A) ($\sigma_{bY_h}^2$ and $\sigma_{wY_h}^2$ for each group), (H+1) for Model (6.2B) ($\sigma_{bY_h}^2$ for each group and a common σ_{wY}^2), and 2 for Model (6.2C) (common σ_{bY}^2 and σ_{wY}^2).

Since the goal of modeling exposure data is to select the most parsimonious model (the one with the fewest parameters) consistent with the data, *likelihood ratio tests* can be applied to compare different versions of Model (6.2). These tests compare the reduced Models (6.2B) and (6.2C) to the full Model (6.2A) to examine the effects of assuming homogeneity of $\sigma_{wY_h}^2$ and/or $\sigma_{bY_h}^2$ across groups. The likelihood ratio test, comparing Model (6.2A) to Model (6.2B), is testing the null hypothesis H_0: $\sigma_{wY_1}^2 = ... = \sigma_{wY_H}^2 = \sigma_{wY}^2$; and, for comparing Model (6.2A) to Model (6.2C), the likelihood ratio test is testing the null hypothesis H_0: $\sigma_{bY_h}^2 = \sigma_{bY}^2$ and $\sigma_{wY_h}^2 = \sigma_{wY}^2$ for all h, h = 1, 2, ... H.

Let L_F and L_R represent the full and reduced log likelihoods (available, for example, from the Proc MIXED output in SAS), respectively, obtained by fitting the full Model (6.2A) and either the reduced Model (6.2B) or the reduced model (6.2C) to a set of data. We define the likelihood ratio statistic as $LR = -2(L_R - L_F)$. Let $\hat{\sigma}_{wY_h}^2$ and $\hat{\sigma}_{bY_h}^2$ (for h = 1, 2, ..., H), and $\hat{\sigma}_{wY}^2$ and $\hat{\sigma}_{bY}^2$, represent the within-person and between-person estimated variance components under the full and reduced models, respectively. Then the following values are assigned to the estimated variance components, depending upon the value of LR, using a Type I error rate of α_{LR} based upon a Chi-square distribution with either $(H\text{-}1)$ d.f. [for comparing Model (6.2A) to (6.2B)] or $2(H\text{-}1)$ d.f. [for comparing Model (6.2A) to (6.2C)]:

$$\hat{\sigma}^2_{wY_h} = \hat{\sigma}^2_{wY_h} \text{ if } LR > \chi^2_{d.f.,1-\alpha_{LR}},$$

$$\hat{\sigma}^2_{wY_h} = \hat{\sigma}^2_{wY} \text{ if } LR \leq \chi^2_{d.f.,1-\alpha_{LR}},$$

$$\hat{\sigma}^2_{bY_h} = \hat{\sigma}^2_{bY_h} \text{ if } LR > \chi^2_{d.f.,1-\alpha_{LR}}, \text{ and}$$

$$\hat{\sigma}^2_{bY_h} = \hat{\sigma}^2_{bY} \text{ if } LR \leq \chi^2_{d.f.,1-\alpha_{LR}}.$$

Based upon simulation studies comparing Model (6.2A) to (6.2B), Weaver *et al.* (2001) recommended conducting the LR test at a significance level of $\alpha_{LR} = 0.01$.

6.3.4 Combining data from multiple groups

A likelihood ratio test was performed on the logarithms of the observed exposure levels, i.e., the $\{y_{hij}\}$, to determine whether it would be legitimate to pool either the within-person, or both the within-person and between-person, variance components from Groups 1 - 4. The results are shown in Table 6.5, using the values of $-2LL$ ('minus two times the log-likelihood') obtained from output of Proc MIXED. The table shows that the null hypotheses of a common within-person variance component [Model (6.2A) vs. Model (6.2B)], or of common within-person and between-person variance components [Model (6.2A) vs. Model (6.2C)], would both be rejected with a P-value <0.001. Results from these likelihood ratio tests indicate that it would be inappropriate to pool these variance components for statistical modeling.

Table 6.5 Likelihood ratio tests comparing the full Model (6.2A) to the reduced Models (6.2B) and (6.2C) for the logged exposure levels from Groups 1 - 4.

Model	$-2LL$*	Model (6.2A) vs. (6.2B)			Model (6.2A) vs. (6.2C)		
		LR	d.f. $(H\text{-}1)$	P-value	LR	d.f. $2(H\text{-}1)$	P-value
6.2A	991.830	74.529	3	<0.001	140.05	6	<0.001
6.2B	1066.359						
6.2C	1131.877						

* Minus two times the log-likelihood from output of SAS Proc MIXED.

Although it appears inappropriate to assign common between-person and/or within-person variance components to Groups 1 – 4, inspection of Table 6.2 suggests that this is largely due to the extremely large estimated variance components for Group 3. Indeed, it may be possible to exclude Group 3 and then assign common between-person and/or within-person variance components for statistical modeling of Groups 1, 2, and 4. Table 6.6 shows results from application of the likelihood ratio tests to the $\{y_{hij}\}$ from Groups 1,

2, and 4. The results indicate that it would be appropriate to apply reduced Model (6.2B), which assumes a common σ_{wY}^2 for Groups 1, 2, and 4 [P-value = 0.237 for the likelihood ratio test comparing Model (6.2A) to Model (6.2B)]. However, it would probably be inappropriate to accept the fully reduced Model (6.2C), which assumes homogeneous σ_{wY}^2 and σ_{bY}^2 for Groups 1, 2, and 4 [P-value = 0.019 for the likelihood ratio test comparing Model (6.2A) to Model (6.2C)]. The estimated parameters obtained from application of Model (6.2B) to the logged exposure levels for Groups 1, 2, and 4 are summarized in Table 6.7.

Table 6.6 Likelihood ratio tests comparing the full Model (6.2A) to the reduced Models (6.2B) and (6.2C) for logged exposure levels from Groups 1, 2, and 4.

Model	-2LL*	Model (6.2A) vs. (6.2B)			Model (6.2A) vs. (6.2C)		
		LR	d.f. (H-1)	P-value	LR	d.f. 2(H-1)	P-value
6.2A	781.197	2.876	2	0.237	11.759	4	0.019
6.2B	784.073						
6.2C	792.957						

* Minus two times the log-likelihood from output of SAS Proc MIXED.

Table 6.7 Parameters estimated under Model (6.2B) for logged exposure levels from Groups 1, 2 and 4.

Group	$\hat{\mu}_{Y_h}$	$\hat{\sigma}_{bY_h}^2$	$\hat{\sigma}_{wY}^2$
1	2.73	0.0212	0.429
2	-0.422	0.0447	0.429
4	4.62	0.296	0.429

6.4 This chapter and Chapter 7

In this chapter, we introduced the flexible family of general linear mixed models and showed how it can be used to efficiently describe chemical exposure data. We explored applications of such models to data from our four test groups and evaluated the underlying statistical assumptions to make sure that the models can be used to make valid inferences about exposure levels. In Chapter 7, we will use the same models [Models (6.2) and (6.3)] to evaluate the effects of covariates on levels of welding-fume exposures in the construction industry, and thereby identify some important determinants of such exposure levels.

7 DETERMINANTS OF EXPOSURE LEVELS

A great strength of Model (6.3) is its ability to quantify important effects of covariates on levels of exposure, while simultaneously providing information about the components of variance within and between persons. As such, Model (6.3) is useful for identifying *determinants of exposure* in occupational and environmental settings (Bakke *et al.*, 2002; Burstyn *et al.*, 2000a; Burstyn *et al.*, 2000b; Burstyn and Teschke, 1999; de Cock *et al.*, 1998; Lumens and Spee, 2001). Our ability to identify such exposure determinants is central to the ultimate control of exposures, as well as for the valid and precise assignment of exposure levels to subjects in epidemiologic studies.

The utility of mixed models will now be illustrated for describing exposure levels of welding fumes among 4 observational groups of construction workers that were identified in Chapter 3 (Section 3.1.1). These data consist of 195 personal measurements of welding fume concentrations among 62 workers in four observational groups, namely, boiler makers (BM), iron workers (IW), pipe fitters (PF), and welder fitters (WF)[13]. Hereafter, we will refer to these four groups as Groups 5 – 8, respectively.

A preliminary exploration of the data is summarized in Figure 7.1 in the form of box and whisker plots of the $\{Y_{hij} = \ln(X_{hij})\}$ for the four groups. These plots point to large differences in average levels of exposure among the groups in the order BM > IW > PF $\cong$ WF. The four distributions of (log-transformed) measurements seem to be roughly normal in shape (indicated by the locations of the mean (+) and median values of the $\{Y_{hij}\}$). Two extremely small exposure values among the welder fitters (from subjects 51 and 59) might be possible outliers.

7.1 Preliminary investigation

Measurements of exposure were obtained from these 62 construction workers at 9 different work sites. Several covariates, related to the process and

---

[13] These data were summarized as 'total particulates' in two published papers (Rappaport *et al.*, 1999; Weaver *et al.*, 2001). The original dataset consisted of 198 observations from 60 workers. However, a coding error resulted in three duplicate observations; thus, only 195 valid measurements will be examined here. Note also that two workers were engaged as pipe fitters on some days and as welder fitters on other days. Thus, for these analyses, the total number of persons in the four groups is taken to be 62.

environment, were recorded on the days of measurement. These covariates were dichotomized for analysis as summarized in Table 7.1.

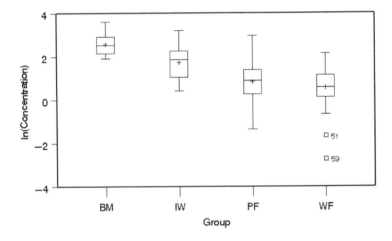

Fig. 7.1 Box and whisker plots of log-transformed air concentrations (mg/m^3) of welding fumes from four groups of construction workers. Legend: BM, boiler makers; IW, iron workers; PF, pipe fitters; WF, welder fitters.

Table 7.1 Covariates recorded with welding-fume exposures for Groups 5 – 8.

Covariate	Code	Values
Work site	*Site*	1 - 9
Type of work	*TW*	1: Welding
		0: Brazing & thermal cutting
Indoor/Outdoor	*IO*	1: Indoor work
		0: Outdoor work
Continuous/Intermittent	*CI*	1: ≤ 50% Hot work
		0: > 50% Hot work
Ventilation	*VE*	1: Local exhaust or mechanical
		0: Natural

Univariate analyses were conducted to investigate the distributions of exposure measurements across covariate values and to determine whether some covariates significantly affected exposure levels. As shown in Table 7.2, data from some groups were specific to particular work sites; in fact, all measurements of BM exposures were obtained from Site 1. Within groups, median exposure levels were significantly different (via the score test available with Proc NPAR1WAY of SAS) across work sites for IW and PF (P-value < 0.05 for both comparisons), but not for WF. When median exposures were compared for each group-covariate combination, some significant differences

Determinants of Exposure

emerged, as shown in Table 7.3. That is, PF workers were exposed to significantly higher levels when $CI = 0$ (indicating >50% hot work), and when $IO = 1$ (indicating indoor work). For all four groups combined, persons were exposed to significantly higher levels when $CI = 0$ (indicating >50% hot work) and to marginally higher levels when $VE = 0$ (indicating natural ventilation).

Table 7.2 Median exposure levels (mg/m^3) and numbers of observations (in parentheses) for each group by work site.

Group(s)	Site								
	1	2	3	4	5	6	7	8	9
5 (BM)	12.5 (20)								
6 (IW)[a]			2.47 (14)			7.81 (27)			
7 (PF)[a]		4.18 (24)		1.74 (25)	2.21 (19)		3.65 (2)		0.784 (2)
8 (WF)							1.52 (22)	1.84 (23)	2.61 (17)
5 – 8[b]	12.5 (20)	4.18 (24)	2.47 (14)	1.74 (25)	2.21 (19)	7.81 (27)	1.58 (24)	1.84 (23)	1.91 (19)

[a] $0.01 \leq P$-value < 0.05 for differences among median exposure levels across work sites.
[b] P-value < 0.01 for differences among median exposure levels across work sites.

Table 7.3 Median exposure levels (mg/m^3) and numbers of observations (in parentheses) for all covariates by group.

Group(s)	TW		CI		VE		IO	
	1	0	1	0	1	0	1	0
5 (BM)	12.5 (18)	31.2 (1)	9.45 (2)	13.0 (17)	12.6 (17)	20.6 (2)	12.8 (2)	10.6 (18)
6 (IW)	7.24 (32)	8.19 (3)	8.44 (8)	5.77 (27)	(0)	7.43 (35)	2.55 (7)	7.62 (28)
7 (PF)	2.50 (48)	1.81 (5)	1.93[b] (17)	3.24 (35)	2.50 (16)	2.50 (37)	3.68[c] (29)	2.15 (24)
8 (WF)	1.88 (42)	1.55 (18)	1.61 (39)	2.14 (23)	1.64 (30)	2.22 (32)	1.58 (31)	1.84 (31)
5-8	3.49 (140)	2.88 (27)	2.50[b] (85)	4.06 (83)	2.49[a] (63)	3.63 (106)	3.66 (85)	2.62 (85)

[a] $0.05 \leq P$-value < 0.10 for differences between median exposure levels in group-covariate combinations.
[b] $0.01 \leq P$-value < 0.05 for differences between median exposure levels in group-covariate combinations.
[c] P-value < 0.01 for differences between median exposure levels in group-covariate combinations.

7.2 Profile plots

Profile plots were constructed for the repeated measurements obtained from all workers in each group, as shown in Figure 7.2. Because all workers were not measured at the same equally-spaced points in time, these plots should be interpreted with caution. Nonetheless, the plots indicate that exposure levels were highly variable within workers over time and point to a possible increasing trend in exposure for IW (Group 6).

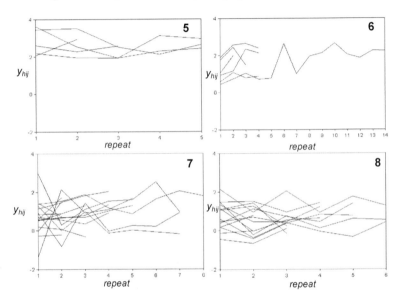

Fig. 7.2. Profile plots of welding fume levels for four groups of construction workers. Each sequence $\{y_{hij}\}$ represents the log-transformed concentration (mg/m^3) repeatedly measured for a given worker in a given group on different days. Numbers refer to groups (5 = BM, 6 = IW, 7 = PF, 8 = WF).

7.3 Application of mixed models

Model (6.2) was applied to the data from Groups 5 – 8 via Proc MIXED of SAS with group mean (μ_{Y_h}) as a fixed effect and person (b_{hi}) as a random effect. Both the full model [Model (6.2A)] and reduced models [Models (6.2B) and (6.2C)] were used. Table 7.4 lists the estimated group mean exposures (values of $\hat{\mu}_{Y_h}$ and $\hat{\mu}_{X_h} = e^{[\hat{\mu}_{Y_h} + 0.5(\hat{\sigma}^2_{bY_h} + \hat{\sigma}^2_{wY_h})]}$), as well as the estimated within-person and between-person variance components ($\hat{\sigma}^2_{wY_h}$ and $\hat{\sigma}^2_{bY_h}$) obtained under Models (6.2A), (6.2B), and (6.2C). Although the values of $\hat{\mu}_{Y_h}$ for a given group were similar under Models (6.2A), (6.2B), and (6.2C), suggesting that the choice of model had little impact upon estimation of the fixed group effects, the estimated variance components differed substantially across the three models, with Model (6.2C) (common within-person and between-person variance components) being the most deviant. These differences were reinforced by results of likelihood ratio tests indicating that it would be legitimate to pool the within-person variance components across the four groups [Model (6.2B) vs. (6.2A), P-value = 0.143], but probably not the between-person variance components [Model (6.2C) vs. (6.2A), P-value = 0.020]. Relying upon the values of $\hat{\mu}_{Y_h}$, $\hat{\sigma}^2_{bY_h}$, and $\hat{\sigma}^2_{wY_h}$, obtained under Model (6.2B), the impact of group differences on exposure levels can be gauged by the 6-fold range of estimated mean exposures (i.e., the values of $\hat{\mu}_{X_h}$) across the four groups.

Table 7.4 Estimated parameters for Models (6.2A) (6.2B) and (6.2C) fit to exposure data for Groups 5 – 8.

Group	$\hat{\sigma}^2_{bY_h}$			$\hat{\sigma}^2_{wY_h}$			$\hat{\mu}_{Y_h}$			$\hat{\mu}_{X_h}$		
	6.2A	6.2B	6.2C	6.2A	6.2B	6.2C	6.2A	6.2B	6.2C	6.2A	6.2B	6.2C
5	0.000	0.000	0.244	0.279	0.362	0.389	2.58	2.58	2.57	15.2	15.8	17.9
6	0.131	0.152	0.244	0.417	0.362	0.389	1.82	1.84	1.85	8.12	8.14	8.73
7	0.148	0.166	0.244	0.449	0.362	0.389	0.799	0.791	0.783	3.00	2.87	3.00
8	0.866	0.712	0.244	0.234	0.362	0.389	0.406	0.424	0.472	2.60	2.61	2.20

Likelihood ratio test: 6.2B vs. 6.2A, P-value = 0.143; 6.2C vs. 6.2A, P-value = 0.020

7.3.1 Normality of standardized random effects and residuals

Standardized random effects were generated under Model (6.2B) for each group (except for BM where the estimated between-person variance component was zero). These estimated random effects divided by their estimated standard errors are plotted in q-q format in Figure 7.4. In each case, the deviations from normality do not seem too great. This impression is reinforced by application of the Shapiro-Wilks W test to these standardized random effects, which suggests no significant deviations from normality (IW: P-value = 0.226; PF: P-value = 0.097; WF: P-value = 0.844). Likewise, as shown in Figure 7.5, q-q lots of the residuals under Model (6.2B) are reasonably normally distributed (Shapiro-Wilks W test, BM: P-value = 0.106; IW: P-value = 0.291; PF: P-value = 0.068; WF: P-value = 0.653), aside from one possible (low) outlier in the PF category (Group 7).

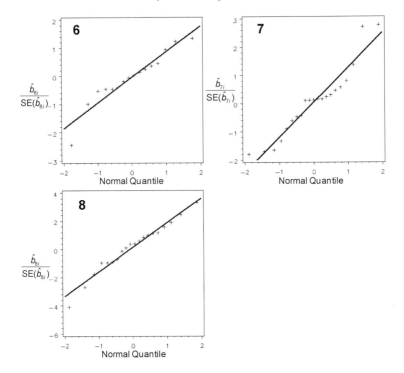

Fig. 7.4 Quantile-quantile plots of standardized random effects obtained from Model (6.2B) for Groups 6-8. In each plot, $\hat{b}_{hi}/\text{SE}(\hat{b}_{hi})$ is the estimated random effect for the ith subject in Group h divided by its estimated standard error, $h = 6, 7, 8$. Numbers refer to groups (6 = IW, 7 = PF, 8 = WF).

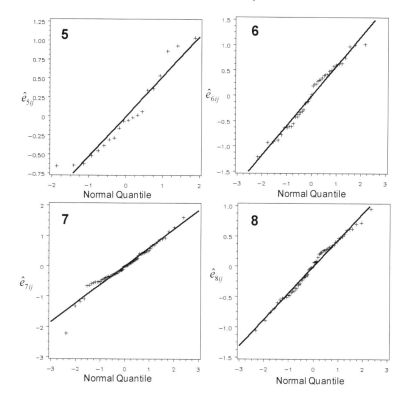

Fig. 7.5 Quantile-quantile plots of residuals obtained from Model (6.2B) for Groups 5-8. In each plot, the residual ($\hat{e}_{hij}$) represents the deviation of the jth observation from its predicted value for the ith person in Group h, $h = 5, 6, 7, 8$. Numbers refer to groups (5 = BM, 6 = IW, 7 = PF, 8 = WF).

7.4 Investigating covariates

Model (6.3B) was employed via Proc MIXED of SAS to investigate the effects of all four dichotomous covariates (*TW, IO, CI,* and *VE*) and two-way interactions, assuming a common σ_{wY}^2 and a distinct $\sigma_{bY_h}^2$ for each group. The total number of fixed effects to be estimated was 26 [8 main effects (4 group means for the BM, IW, PF, and WF groups, plus *IO, CI, VE,* and *TW*), 12 group-covariate interactions, and 6 pairwise covariate interactions]. A backward elimination strategy was used to remove non-significant (*P*-value > 0.10) interaction effects. This led to the final regression model shown in Table 7.5, containing 8 main effects and 4 interaction terms.

Table 7.5 Final Model (6.3B) for welding fume exposure concentrations (effects estimated assuming common within-person, and distinct between-person, variance components).

Predictor	Estimate	P-value
BM (Group 5)	1.486	0.0281
IW (Group 6)	1.230	0.0003
PF (Group 7)	0.0226	0.9428
WF (Group 8)	0.2336	0.4468
CI	0.2261	0.1936
VE	-0.0369	0.8505
IO	0.7471	0.0194
TW	0.5843	0.0201
(BM)(IO)	1.178	0.0752
(PF)(IO)	1.178	0.0005
(CI)(VE)	-0.5400	0.0337
(IO)(TW)	-1.3828	<0.0001
Variance Components	Estimate	
BM: ($\hat{\sigma}_{bY_5}^2$)	0	
IW: ($\hat{\sigma}_{bY_6}^2$)	0	
PF: ($\hat{\sigma}_{bY_7}^2$)	0.1092	
WF: ($\hat{\sigma}_{bY_8}^2$)	0.4172	
Within-Person ($\hat{\sigma}_{wY}^2$)	0.3150	

The residuals from this model were investigated for normality, both collectively and on a group-by-group basis, and showed no evidence of significant lack of fit (results not shown). Likewise, q-q plots of the standardized estimated random effects, using the PF and WF data, suggested no meaningful evidence of deviations from normality. (Note that random effects could not be estimated using BM and IW data because the estimated between-person variance components were zero).

7.4.1 Exposures predicted from fixed effects

The estimated regression coefficients, shown in Table 7.5, were used to predict mean exposures (shown in Table 7.6) for various combinations of predictor values. These results illustrate the influence of pairs of activities (namely, IO and TW) and controls [namely, $CI = 1$ (an administrative control) and $VE = 1$ (an engineering control)]. Comparing categories of work activities and controls within groups, mean exposures varied by about 3-fold to 10-fold. In all groups, the highest predicted mean exposure was for indoor brazing and cutting without controls. However, the lowest predicted mean exposures were not always observed in the same category for the four groups. In two of the four groups (BM and PF), the lowest mean exposure was predicted for outdoor brazing and cutting with controls. But, for IW and WF, the lowest mean exposure was predicted for indoor welding with controls. These apparent anomalies were due to the interaction effects between indoor work ($IO = 1$) and group that were observed for boiler makers ($BM = 1$) and pipe fitters ($PF = 1$) but not for iron workers ($IW = 1$) and welder fitters ($WF = 1$). These interactions offset the reduction in fume levels associated with indoor welding ($IO = TW = 1$), compared to indoor brazing and thermal cutting ($IO = 1$, $TW = 0$), among iron workers and welder fitters, and point to unexplained differences involving indoor work across groups. Such differences in welding fume levels across jobs can point to useful avenues for controlling exposures, as will be discussed in Chapter 9.

Table 7.6 Predicted mean exposure levels for various combinations of predictor values using Model (6.3B), summarized in Table 7.5.

Group	Activity	Exposure without Controls* ($\hat{\mu}_{X_h}$, mg/m³)	Exposure with Controls* ($\hat{\mu}_{X_h}$, mg/m³)	Fold Range (High/Low)
BM	Outdoor brazing/cutting	5.17	3.64 (Low)	9.8
BM	Outdoor welding	9.28	6.53	
BM	Indoor brazing/cutting	35.5 (High)	25.0	
BM	Indoor welding	16.0	11.2	
IW	Outdoor brazing/cutting	4.01	2.82	3.2
IW	Outdoor welding	7.19	5.06	
IW	Indoor brazing/cutting	8.46 (High)	5.95	
IW	Indoor welding	3.81	2.68 (Low)	
PF	Outdoor brazing/cutting	1.20	0.843 (Low)	9.7
PF	Outdoor welding	2.15	1.51	
PF	Indoor brazing/cutting	8.21 (High)	5.78	
PF	Indoor welding	3.70	2.60	
WF	Outdoor brazing/cutting	1.48	1.04	3.2
WF	Outdoor welding	2.65	1.87	
WF	Indoor brazing/cutting	3.12 (High)	2.20	
WF	Indoor welding	1.40	0.989 (Low)	

* Controls consisted of local-exhaust or mechanical ventilation ($VE = 1$) and reduction of hot work to less than 50% of the work day ($CI = 1$). Thus, 'without controls' means $CI = VE = 0$, and 'with controls' means $CI = VE = 1$.

The effects of the various categories of work activities and controls are further illustrated in Figure 7.6. For all 4 groups, the presence of controls ($CI = VE = 1$) significantly reduced exposures in the expected directions. However, the magnitudes and patterns of the reductions across categories differed according to the particular job groups. As noted above, indoor effects were accentuated for boiler makers and pipe fitters compared to iron workers and welder fitters.

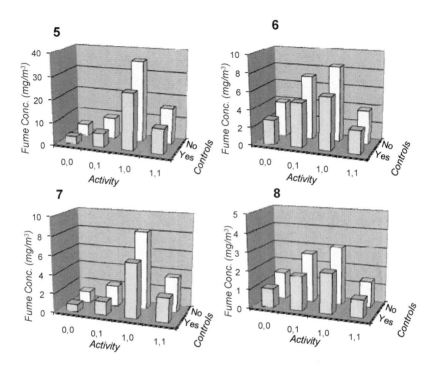

Fig. 7.6 Predicted mean exposures to welding fumes for Groups 5 - 8 (5=BM; 6=IW; 7=PF; 8=WF), based upon the model shown in Table 7.5. Activity (*IO, TW*): (0,0)=outdoor brazing/cutting; (0,1)=outdoor welding; (1,0)=indoor brazing/cutting; (1,1)=indoor welding. Controls consisted of local-exhaust or mechanical ventilation (*VE* = 1) and reduction of hot work to less than 50% (*CI* = 1). (Note that magnitudes of the y-axes differ across groups).

7.4.2 Modeling the random effects

The estimated within-person and between-person variance components, obtained under Model (6.2B) and Model (6.3B), are compared in Table 7.7. When covariates are considered [Model (6.3B)], both the estimated within-person and between-person variance components were reduced in value relative to corresponding estimates under Model (6.2B). While $\hat{\sigma}^2_{wY}$ diminished about 13% (from 0.362 to 0.315), the major impact of adding covariates to the model was to reduce $\hat{\sigma}^2_{bY_h}$ for the three groups with nonzero estimates; i.e., $\hat{\sigma}^2_{bY_h}$ was reduced from 0.152 to zero for IW (100% reduction), from 0.166 to 0.109 for PF (34% reduction), and from 0.712 to 0.417 for WF (41% reduction). This suggests that much of the between-person variability in these groups arose from the unequal distribution of the designated activities and the presence of controls across the workers, a finding consistent with reports from other investigations (Bakke *et al.*, 2002; Burstyn and Kromhout, 2000; Peretz *et al.*, 2002).

Table 7.7. Variance components estimated for Groups 5-8 under Models (6.2B) and (6.3B).

Group	$\hat{\sigma}^2_{bY_h}$		$\hat{\sigma}^2_{wY}$	
	Model (6.2B)	Model (6.3B)	Model (6.2B)	Model (6.3B)
5 (BM)	0.000	0.000	0.362	0.315
6 (IW)	0.152	0.000	0.362	0.315
7 (PF)	0.166	0.109	0.362	0.315
8 (WF)	0.712	0.417	0.362	0.315

7.5 Conclusions

The results illustrated in Tables 7.6 and 7.7 and Figure 7.6 indicate that particular sets of work activities and controls explained significant amounts of the variability in welding-fume exposure levels experienced by construction workers. The observed effects indicate that recognized engineering and administrative controls ($VE = CI = 1$) led to reduced exposures to welding fumes (Figure 7.6). Also, the addition of the covariates to the mixed model meaningfully reduced the magnitudes of the estimated between-person variance components in the groups, suggesting that much of the heterogeneity in exposure concentrations within a group was due to the uneven distribution of work activities and controls across workers.

Certainly, the imbalance of data among particular cells (illustrated in Table 7.3) could have led to unreliable modeling results, and future studies should seek to have more balanced designs. Nonetheless, this case study points to the strength of linear mixed models for characterizing variability of exposure levels arising from both deterministic (fixed effects of different groups and covariates) and random sources.

7.6 This chapter and Chapter 8

In this chapter, we used linear mixed Models (6.2) and (6.3) to characterize construction workers' exposures to welding fumes in four observational groups and to identify a pair of controls (one administrative, one engineering) that significantly reduced fume levels in these groups. In Chapter 8, we will show how the within-person and between-person variance components estimated with linear mixed models can be used to estimate probabilities that exposure levels exceed OELs. We will also relate these probabilities to health risks posed by such exposures.

8 PROBABILITIES OF EXCEEDING OELs

Governmental inspectors, such as those from OSHA, rarely assess workers' exposures to toxic chemicals in the U.S.[14] Rather, the vast majority of occupational exposures are assessed by employers, who are legally obliged to provide a workplace "... free from recognized hazards..." (OSH-Act, 1970). In doing so, a representative of the employer (usually an occupational hygienist) measures personal air levels for one or more workers in each observational group where exposure is thought to be potentially excessive. Having collected a few air samples, the employer decides whether exposure levels are acceptable for each group. In making these decisions, measurements are generally compared one-to-one with the OEL; if all observed exposure concentrations for a group fall below the OEL, then *compliance* is declared and no additional monitoring is required. If, on the other hand, at least one exposure measurement exceeds the OEL, then the employer is required to reduce air concentrations for that group. This direct comparison of air measurements with the OEL has been termed *compliance testing* for obvious reasons (Rappaport, 1991b), and represents the primary tool for managing occupational health risks from chemical agents in the U.S. Even in countries where the legal basis for monitoring may be unclear, compliance testing is commonly practiced because of its seductive simplicity.

8.1 Pitfalls of compliance testing

It has been argued that compliance testing imposes disincentives upon employers to monitor exposures (Rappaport, 1984). To illustrate this point, we first define the *exceedance* of the hth group (γ_h) as the probability that a typical person in Group h would be exposed to an air level above the OEL on any randomly selected day. That is,

$$\gamma_h = P\{X_{hij} > \text{OEL}\} = P\{Y_{hij} > \ln(\text{OEL})\}, \tag{8.1}$$

where X_{hij} is the jth exposure measurement for subject i in Group h. The corresponding probability of compliance for Group h, designated $P\{C_h\}$, is given by

$$P\{C_h\} = (1 - \gamma_h)^{N_h} \tag{8.2}$$

---

[14] Fewer than 10,000 'health' inspections, which can include collection of air samples, are conducted by OSHA each year. There are roughly 2.5 million workplaces in the U.S.

where N_h represents the number of exposure measurements obtained from that group (Rappaport, 1984).[15] This relationship is illustrated in Figure 8.1, which shows the probability of noncompliance (1-P$\{C_h\}$) as a function of γ_h for different sample sizes. Since P$\{C_h\}$ varies inversely with N_h, the employer has an implicit incentive to maximize P$\{C_h\}$ by making very few measurements (that is, P$\{C_h\}$ monotonically approaches 1 as N_h decreases toward zero). Since $0 < \gamma_h \leq 1$, Equation (8.2) also makes explicit what is known intuitively, that any group can be declared out of compliance with very high probability given a large enough sample. Indeed, Figure 8.1 shows that sample size is often the most important single determinant of compliance. Since N_h is totally unrelated to any health hazard, this illustrates a critical flaw in compliance testing.

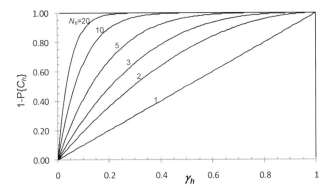

Fig. 8.1. Relationship between the exceedance (γ_h) and the probability of noncompliance (1-P$\{C_h\}$) for Group h based upon sample sizes (N_h) between one and twenty measurements [from Equation (8.2)].

The enormous variability in air levels makes it likely that compliance testing, with small sample sizes, will lead to errors in assessing whether or not a particular work environment poses a hazard to the health of workers. In fact, given the positive skewness of exposure distributions observed in workplaces, situations would be expected where large health risks exist despite a declaration of compliance. To address this issue, we will next explore the long-term risks of disease inherent in OSHA's new standards.

---

[15] Equation (8.2) implicitly assumes that the N_h exposure measurements are mutually independent; this would require that only one measurement be collected per randomly chosen person for the determination of compliance.

8.2 Risk and overexposure

As discussed in Chapter 2, there is evidence that OSHA establishes a new OEL[16] as the lowest concentration that can feasibly be achieved in the worst segment(s) of industries affected by the standard (Rappaport, 1993b). In doing so, OSHA assumes that workers' lifetime risks are proportional to the corresponding cumulative exposures over 45 years. Let $CE_{hi}(t)$ represent the cumulative exposure received by person i in Group h after t years. Then, each OEL imposes the consistent restriction that $CE_{hi}(t) \leq$ OEL(45 years), $t \leq 45$ years. Thus, the health risk in Group h would exceed the allowable risk only when $CE_{hi}(t) >$ OEL(45 years), a condition we will refer to as *overexposure*. By assuming that all persons in the group are ultimately exposed to a contaminant over the same period of $t = 45$ years, the *probability of overexposure* in Group h (θ_h) is given by

$$\begin{aligned} \theta_h &= P\{CE_{hi}(45 \text{ years}) > \text{OEL}(45 \text{ years})\} \\ &= P\{\mu_{X_{hi}}(45 \text{ years}) > \text{OEL}(45 \text{ years})\} \\ &= P\{\mu_{X_{hi}} > \text{OEL}\}, \end{aligned} \qquad (8.3)$$

where $\mu_{X_{hi}}$ is the mean exposure level for a typical person in Group h and $CE_{hi}(t)$ is equal to $\mu_{X_{hi}}$ (45 years). While Equation (8.3) follows logically from OSHA's procedures for setting new standards (i.e., when $t = 45$ years), it is also supported on theoretical and empirical grounds. For example, it is well accepted that the long-term doses of chemicals and their metabolites tend to be proportional to $CE_{hi}(t)$ when exposures are in the low-dose linear range (Hattis, 1998; Lutz, 1998; Olson and Cumming, 1981; Rappaport, 1991b; Rappaport, 1993a). More recent work indicates that the long-term dose received by person i in group h is also proportional to $CE_{hi}(t)$ even when exposure levels extend into the nonlinear range of saturable metabolism (Rappaport et al., 2005). However, it must be emphasized that θ_h relates to the risk due to long-term exposures and not to acute effects associated with exposures of one day or less or to allergenic or reproductive effects, where the timing as well as the intensity of exposure can be important.

The notion that exposure should be evaluated in terms of the mean air concentration inhaled over time has been around for some time. It was argued in Great Britain in the early 1950s that, because of the cumulative effect of coal dust on respiratory function, exposure to coal dust should be assessed in terms of the long-term mean exposure (Long, 1953; Oldham and Roach, 1952; Roach, 1953; Wright, 1953). Somewhat later, attention shifted to statistical methods to estimate means of lognormal exposure distributions (Coenen, 1966;

---

[16] Recall from Chapter 2 that OSHA refers to its OELs as Permissible Exposure Limits (PELs) or Action Levels (ALs). Here, we use the generic designation 'OEL' for simplicity.

Coenen, 1971; Juda and Budzinski, 1964; Juda and Budzinski, 1967; Tomlinson, 1957) and to the consideration of exposures to radioactive particles, for which cumulative exposure is also important [e.g., Sherwood (1966); Langmead (1970)]. Since these early papers were based on air measurements prior to the widespread use of personal sampling, it is unclear whether cumulative exposure was considered relative to the group or to the individual person. However, based upon our current knowledge that mean exposure concentrations can vary considerably among members of an observational group, it should be clear that some persons can be overexposed even when the group mean exposure level is less than the OEL. Certainly, modern epidemiologic studies consider individual cumulative exposures as the most relevant predictors of long-term risks of disease [e.g., see Steenland and Deddens (2004)].

Because of the clear connection between the cumulative exposure and the long-term risk of disease, we regard θ_h (which involves cumulative exposure) as a *gold standard* with which to manage workplace risks from chemical exposure. In Chapter 9, we will present a statistical inference strategy for relating θ_h to OELs and for optimizing options for reducing unacceptable exposures. In the remaining sections of the current chapter, we will more fully explore the interrelationships among θ_h, γ_h, $P\{C_h\}$ and group mean exposures, and we will address associated implications for managing workplace risks.

8.3 Relating probabilities to exposure distributions

From Equations (8.2) and (8.3), we see the potential for difficulties when exposure assessment focuses upon compliance (related to γ_h and N_h) rather than upon the probability of overexposure (related to $\mu_{X_{hi}}$). In order to quantify the health risks and $P\{C_h\}$ in various occupational groups, it is necessary first to relate θ_h and γ_h to the parameters of the underlying distributions of exposure (Spear and Selvin, 1989; Tornero-Velez et al., 1997). Since $Y_{hij} \sim N(\mu_{Y_h}, \sigma_{bY_h}^2 + \sigma_{wY_h}^2)$ based on Model (6.2), it follows from Equation (8.1) that the exceedance of Group h is given by:

$$\gamma_h = 1 - \Phi\left\{\frac{\ln(\text{OEL}) - \mu_{Y_h}}{\sqrt{\sigma_{bY_h}^2 + \sigma_{wY_h}^2}}\right\}, \tag{8.4}$$

where $\Phi\{z\}$ denotes the probability that a standard normal variate Z would take a value no greater than z. Since $\ln(\mu_{X_{hi}}) \sim N(\mu_{Y_h} + \frac{\sigma_{wY_h}^2}{2}, \sigma_{bY_h}^2)$, it follows that the corresponding probability of overexposure is:

$$\theta_h = 1 - \Phi\left\{\frac{\ln(\text{OEL}) - \mu_{Y_h} - \frac{\sigma^2_{wY_h}}{2}}{\sqrt{\sigma^2_{bY_h}}}\right\}. \tag{8.5}$$

To illustrate calculations employing Equations (8.4) and (8.5), in Figure 8.2A we extend the hypothetical example given in Chapter 5, where $\mu_{Y_h} = 2.30$, $\sigma^2_{Y_h} = 0.693$, and $\sigma^2_{bY_h} = \sigma^2_{wY_h} = 0.693/2$. Since the hypothetical OEL = 40 (arbitrary units), then $\gamma_h = 1 - \Phi\left\{\frac{\ln(40) - 2.30}{\sqrt{0.693}} = 1.67\right\} = 0.047$, from which we infer about a 5% chance that a randomly selected exposure measurement from Group h would exceed the OEL. Note that this probability refers to the group as a whole (i.e., to a typical group member) because specific group members can have exceedances (values of γ_{hi}) greater or less than 0.047. For this hypothetical group, γ_{hi} varied between 0.001 and 0.203 for the 5 representative subjects. Extending this example to the probability of overexposure, we have

$$\theta_h = 1 - \Phi\left\{\frac{\ln(40) - 2.3 - \frac{0.346}{2}}{\sqrt{0.346}} = 2.07\right\} = 0.020,$$ indicating a 2% probability that

a typical person in Group h would be overexposed (Figure 8.2B). Note that θ_h also relates to the group as a whole because any specific person in the group has a mean exposure $\mu_{X_{hi}}$ either greater or less than the OEL. In fact, each of the 5 hypothetical workers in this example is not overexposed (i.e., $\mu_{X_{hi}} < 40$, $i = 1, 2, 3, 4, 5$).

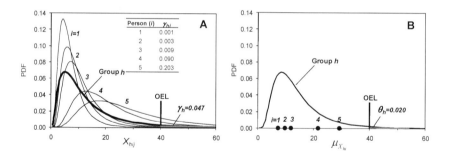

Fig. 8.2 Hypothetical lognormal distributions of exposures for an observational group relative to an OEL of 40 (arbitrary units). A) Probabilities that each of 5 hypothetical persons in Group h ($i = 1, 2, 3, 4, 5$), and that of a typical person in Group h, would be exposed above the OEL on a randomly selected day. (The exceedance for the ith person is γ_{hi} and that for any randomly chosen member of Group h is γ_h). B) Probability that a randomly selected person from Group h would have a mean exposure ($\mu_{X_{hi}}$) greater than the OEL; this is the probability of overexposure, θ_h. The dots represent the values of $\mu_{X_{hi}}$ ($i = 1, 2, 3, 4, 5$) for the five persons in A.

8.3.1 Estimating exceedance and probability of overexposure

Equations (8.4) and (8.5) can be used along with estimated parameters based on available exposure data to estimate γ_h and θ_h. That is,

$$\hat{\gamma}_h = 1 - \Phi\left\{\frac{\ln(\text{OEL}) - \hat{\mu}_{Y_h}}{\sqrt{\hat{\sigma}^2_{bY_h} + \hat{\sigma}^2_{wY_h}}}\right\} \text{ and } \hat{\theta}_h = 1 - \Phi\left\{\frac{\ln(\text{OEL}) - \hat{\mu}_{Y_h} - \frac{\hat{\sigma}^2_{wY_h}}{2}}{\sqrt{\hat{\sigma}^2_{bY_h}}}\right\}.$$ Table 8.1

shows the REML-based estimates of γ_h and θ_h for Groups 1 - 8 (relative to OELs at the time of monitoring). Note that the estimates of the exceedance and the probability of overexposure do not always coincide; that is, $\hat{\theta}_h < \hat{\gamma}_h$ for Groups 1, 2 and 7; $\hat{\theta}_h \cong \hat{\gamma}_h$ for Groups 4 and 8; and $\hat{\theta}_h > \hat{\gamma}_h$ for Groups 3 and 6. The reason for this behavior will be discussed later.

Table 8.1 Estimates of exceedance (γ_h) and the probability of overexposure (θ_h) for Groups 1 – 8. [Based upon Equations (8.4) and (8.5), using REML estimates of the parameters given in Table 6.1].

Group	Contaminant	OEL[a]	$\hat{\gamma}_h$	$\hat{\theta}_h$
1	Inorganic lead	50 µg/m³ (PEL)	0.031	<0.001
		30 µg/m³ (AL)	0.144	0.001
2	Benzene	3.2 mg/m³ (PEL)	0.012	<0.001
		1.6 mg/m³ (AL)	0.103	<0.001
3	Benzene	3.2 mg/m³ (PEL)	0.300	0.577
		1.6 mg/m³ (AL)	0.410	0.738
4	Styrene	213 mg/m³ (PEL)	0.202	0.180
5	Welding fumes	5.0 mg/m³ (TLV)	0.947	Undefined[b]
6	Welding fumes	5.0 mg/m³ (TLV)	0.616	0.842
7	Welding fumes	5.0 mg/m³ (TLV)	0.132	0.061
8	Welding fumes	5.0 mg/m³ (TLV)	0.123	0.113

[a] OELs refer to OSHA's Permissible Exposure Limits (PELs) and Action Levels (ALs, if available) or the ACGIH TLV-TWA (for welding fumes) at the time that measurements were made.
[b] Because the estimated between-person variance component was zero, the probability of overexposure cannot be calculated.

8.3.2 A large sample of observational groups

The estimated values of γ_h and θ_h for Groups 1 – 8 covered the entire range from essentially zero to one (Table 8.1), and $\hat{\gamma}_h$ was generally greater than $\hat{\theta}_h$. To obtain insight into the distributions of $\hat{\gamma}_h$ and $\hat{\theta}_h$ in many workplaces, Tornero-Velez et al. (1997) compiled estimates of γ_h and θ_h from 179 observational groups of workers, with Model (6.2) fit separately to each group.

The results, shown graphically in Figure 8.3A, indicate that $\hat{\theta}_h < \hat{\gamma}_h$ in about 80% of the groups. Overall, groups experienced relatively small probabilities of exceeding the OEL since 73% of $\hat{\gamma}_h$ values and 79% of $\hat{\theta}_h$ values were less than 0.1. Nonetheless, about one group in five had a value of $\hat{\theta}_h > 0.1$, indicative of unacceptable levels of exposure. Also, the cumulative distributions of $\hat{\gamma}_h$ and $\hat{\theta}_h$ intersected when $\hat{\gamma}_h$ was about 0.2; therefore, when exceedance was greater than about 0.20, it was typically smaller than the corresponding probability of overexposure. This behavior is seen more clearly in Figure 8.3B, which is a plot of $\hat{\theta}_h$ versus $\hat{\gamma}_h$ for the same 179 groups. As $\hat{\gamma}_h$ increases above about 0.20, virtually all values of $\hat{\theta}_h$ fell above the 45 degree line representing $\hat{\gamma}_h = \hat{\theta}_h$.

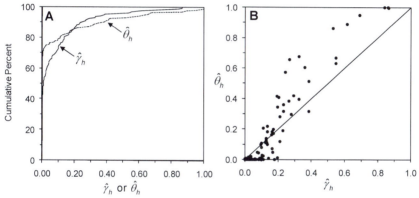

Fig. 8.3 A) Cumulative distributions of estimates of exceedance ($\hat{\gamma}_h$) and the probability of overexposure ($\hat{\theta}_h$) for 179 observational groups of workers. B) $\hat{\theta}_h$ versus $\hat{\gamma}_h$ for the same 179 groups of workers. The 45-degree line represents the situation where $\hat{\theta}_h = \hat{\gamma}_h$. Data are from Tornero-Velez et al. (1997).

8.4 Compliance vs. health risk

We now return to the notion that, due to small sample sizes, compliance testing can misrepresent the health risks posed by long-term chemical exposures. The greatest concern would be that compliance would be declared for a given group of workers despite rather large probabilities of overexposure (θ_h) and associated large, long-term, health risks. Since industrial surveys within a calendar year rarely include more than 4 measurements per group (Tornero-Velez et al., 1997) (i.e., one measurement on each of 4 persons), we illustrate the problem

by assuming that $1 \leq N_h \leq 4$. Using Equations (8.4) and (8.5), with
$z_{1-\gamma_h} = \dfrac{\ln(\text{OEL}) - \mu_{Y_h}}{\sqrt{\sigma_{bY_h}^2 + \sigma_{wY_h}^2}}$ and $z_{1-\theta_h} = \dfrac{\ln(\text{OEL}) - \mu_{Y_h} - \sigma_{wY_h}^2}{\sqrt{\sigma_{bY_h}^2}}$, it follows that

$$z_{1-\theta_h} = \dfrac{z_{1-\gamma_h}\sigma_{Y_h} - (\frac{\sigma_{wY_h}^2}{2})}{\sigma_{bY_h}} = \dfrac{z_{1-\gamma_h} - \frac{\sigma_{Y_h}}{2}(1-\rho_h)}{\sqrt{\rho_h}}, \quad (8.6)$$

where $\rho_h = \dfrac{\sigma_{bY_h}^2}{\sigma_{Y_h}^2}$ is the intraclass correlation for Group h and where $\sigma_{Y_h}^2 = (\sigma_{bY_h}^2 + \sigma_{wY_h}^2)$. From Equations (8.1) and (8.6), the probabilities of compliance (P{C_h}) and overexposure (θ_h) are given in Table 8.2 for various combinations of γ_h, $\sigma_{Y_h}^2$, and ρ_h, representing 50th, 75th and 90th percentiles of the cumulative distributions of the respective estimates $\hat{\gamma}_h$, $\hat{\sigma}_{Y_h}^2$, and $\hat{\rho}_h$ from Tornero-Velez et al. (1997).

Table 8.2. Probabilities of compliance P{C_h} (for sample size N_h ranging from 1 to 4) and the corresponding probabilities of overexposure θ_h for various combinations of γ_h, ρ_h, and $\sigma_{Y_h}^2$ *.

γ_h	ρ_h	P{C_h}	θ_h		
			$\sigma_{Y_h}^2 = 0.52$	$\sigma_{Y_h}^2 = 1.14$	$\sigma_{Y_h}^2 = 1.85$
0.01	0.04	0.96-0.99	0.000	0.000	0.000
0.01	0.22	0.96-0.99	0.000	0.000	0.000
0.01	0.41	0.96-0.99	0.000	0.001	0.001
0.11	0.04	0.63-0.89	0.000	0.000	0.002
0.11	0.22	0.63-0.89	0.022	0.042	0.069
0.11	0.41	0.63-0.89	0.057	0.077	0.099
0.25	0.04	0.32-0.75	0.050	0.209	0.457
0.25	0.22	0.32-0.75	0.201	0.291	0.379
0.25	0.41	0.32-0.75	0.235	0.287	0.335

*Values for γ_h, $\sigma_{Y_h}^2$, and ρ_h represent 50th, 75th, and 90th percentiles of the $\hat{\gamma}_h$, $\hat{\sigma}_{Y_h}^2$ and $\hat{\rho}_h$ distributions, respectively, reported by Tornero-Velez et al. (1997).

The table shows that the probability of overexposure would be small ($\theta_h \leq 0.001$) for the typical case when $\gamma_h = 0.01$, regardless of the particular values of ρ_h and $\sigma_{Y_h}^2$. In such low-risk situations, compliance testing should work well with 4 or fewer measurements, since $0.96 \leq$ P{C_h} ≤ 0.99. However, for high-risk situations (i.e., when $0.11 \leq \gamma_h \leq 0.25$), compliance would still be declared between 32% and 89% of the time (i.e., $0.32 \leq$ P{C_h} ≤ 0.89), despite values of

θ_h as large as 0.457. Thus, while a non-compliance decision can generally be interpreted to imply an unacceptably large health risk, a compliance decision cannot be interpreted as implying an acceptably small health risk.

To illustrate the importance of sample size on compliance testing, consider the data for Groups 1 – 8. Using the estimated values of γ_h, summarized in Table 8.1, the probabilities of noncompliance, i.e., $(1-P\{C_h\})$, can easily be estimated by substituting values of $\hat{\gamma}_h$ and N_h into Equation (8.2). For example, for Group 1 ($h = 1$), $\hat{\gamma}_1 = 0.144$ for the Action Level (AL) of 30 µg/m³, from which $(1-P\{C_1\}) = 0.14, 0.37, 0.54,$ and 0.96 for $N_1 = 1, 3, 5$ and 20, respectively. So, an assessment with at least five mutually independent measurements would more likely than not produce at least one exposure value above the AL; then, additional monitoring would eventually lead to a measurement above the PEL and the concomitant decision of noncompliance (Rappaport, 1984). A decision as to whether exposures for members of Group 1 are in compliance would, therefore, be reduced to the following troubling statement: if an initial assessment included fewer than five randomly selected measurements, then compliance is the likely outcome; and, if the assessment included five or more randomly chosen exposure measurements, then noncompliance is more likely.

8.5 Using exceedance to characterize acceptable exposure

As shown in Tables 8.1 and 8.2 and Figure 8.3B, the exceedance and probability of overexposure are often quite different for a given group. This difference reflects the complicated relationship between γ_h and θ_h, which depends upon the particular values of the within-person and between-person variance components for that group (Spear and Selvin, 1989; Tornero-Velez et al., 1997). In general, $\gamma_h > \theta_h$ when γ_h is 'small', and $\gamma_h < \theta_h$ when γ_h is 'large'. However, what constitutes a 'small' or 'large' value of γ_h depends upon $\sigma^2_{Y_h}$, which, as we have seen, covers a remarkably large range. Thus, for some values of $\sigma^2_{Y_h}$, γ_h is a conservative surrogate for θ_h (i.e., when $\gamma_h > \theta_h$); and, for other values of $\sigma^2_{Y_h}$, γ_h is an anticonservative surrogate for θ_h (when $\gamma_h < \theta_h$).

Several investigators have promoted the assessment of occupational exposures based upon values of γ_h, presuming that this strategy will also limit long-term disease risks (related to θ_h) (AIHA, 1990; Corn and Esmen, 1979; Esmen, 1992; Leidel et al., 1977; Lyles and Kupper, 1996; Selvin et al., 1987; Tuggle, 1982). Since this is not always the case, let us further explore the conditions under which the exceedance γ_h is a conservative surrogate for the probability of overexposure θ_h.

Recalling Equation (8.6), in which we related $z_{1-\theta_h}$ to $z_{1-\gamma_h}$ as

$$z_{1-\theta_h} = \frac{z_{1-\gamma_h}\sigma_{Y_h} - \frac{\sigma_{wY_h}^2}{2}}{\sigma_{bY_h}} = \frac{z_{1-\gamma_h} - \frac{\sigma_{Y_h}}{2}(1-\rho_h)}{\sqrt{\rho_h}},$$ it can be shown that

$$\gamma_h < 1 - \Phi\left\{\frac{\sigma_{Y_h}}{2}\left(1+\sqrt{\rho_h}\right)\right\} \tag{8.7}$$

when $\theta_h < \gamma_h$. Recognizing that $0 \le \rho_h \le 1$, the maximum value of the function inside the brackets in Equation (8.7) occurs when $\rho_h = 1$ (i.e., this maximum value is $\frac{\sigma_{Y_h}}{2}(1+\sqrt{1}) = \sigma_{Y_h}$); and, this leads in turn to the smallest upper bound (namely, $1-\Phi\{\sigma_{Y_h}\}$) for γ_h such that $\theta_h < \gamma_h$ for a given value of $\sigma_{Y_h}^2$, regardless of the magnitude of the intraclass correlation ρ_h. So, for any group having a particular value of $\sigma_{Y_h}^2$, any value of $\gamma_h < 1-\Phi\{\sigma_{Y_h}\}$ would necessarily be greater than θ_h (i.e., γ_h would be a conservative surrogate for θ_h). However, because there is no upper limit that the variance $\sigma_{Y_h}^2$ might have in an *uncharacterized* workplace, there is no theoretical upper bound for γ_h that one would adopt *a priori* as a 'conservative' bound for θ_h in prospective assessments of occupational exposures. Based on data reported by Tornero-Velez et al. (1997) from 179 occupational groups, the 90th percentile value for $\hat{\sigma}_{Y_h}^2$ was 1.84. Using $\hat{\sigma}_{Y_h}^2 = 1.84$ to represent a highly variable distribution of exposure levels, a generally conservative *a priori* value of the exceedance for occupational exposure assessment would be any value of $\gamma_h < 1-\Phi\{\sqrt{1.84}\}$ = 0.087, which is somewhat larger than the value of $\gamma_h = 0.05$ advocated by some professional organizations (AIHA, 1990; Mulhausen and Damiano, 1998). Thus, basing decisions upon an allowable exceedance of $\gamma_h = 0.05$ will likely lead to very conservative (i.e., highly protective) outcomes relative to the long-term risks posed by the exposures. That is, by setting the level of allowable exceedance γ_h at 0.05, say, there will be a set of corresponding probabilities of overexposure (depending upon $\sigma_{Y_h}^2$ and ρ_h), all of which are likely to be smaller than 0.05.

Although statistical methods are available for assessing acceptable exceedance via upper one-sided tolerance limits (Lyles and Kupper, 1996), it is important to understand that *these methods require uncorrelated data* and can only be applied under Model (4.1), where a single measurement is obtained from each of a random sample of persons in an observational group. Furthermore, the number of persons (each with one exposure measurement) required to demonstrate acceptable exposure (based upon one-sided tolerance limits) can be large. To illustrate such sample sizes requirements, Table 8.3 is reproduced from Lyles and Kupper (1996) to show the numbers of workers (k_h) from Group h (each having a single exposure measurement) that would be required to demonstrate that $\gamma_h < 0.05$ with a significance level of $\alpha = 0.05$ and power of $(1-\beta) = 0.80$. Calculations are shown for various values of OEL/$X_{0.95}$, representing the ratio of the OEL to the 95th percentile value of the distribution of X_{hij}, when $\sigma^2_{Y_h}$ ranges from 0.5 to 3.0. The results indicate that many workers would need to be measured to achieve an 80 percent level of power when OEL/$X_{0.95}$ is less than about three. For example, using the estimated 95th percentiles of exposure and variance for Group 2 (OEL = PEL = 3.2 mg/m^3, OEL/$\hat{X}_{0.95}$ = 1.5; $\hat{\sigma}^2_{Y_h}$ = 0.50), it appears that 58 workers would be required to provide statistical evidence of acceptable exposure.

Table 8.3 Number of workers (k_h) required for monitoring (one measurement per worker) to provide statistical evidence that the group exceedance (γ_h) is less than 0.05 with a significance level of $\alpha = 0.05$ and power = 0.80 [from Lyles and Kupper (1996)].

OEL/$X_{0.95}$	$\sigma^2_{Y_h}=0.50$	$\sigma^2_{Y_h}=1.0$	$\sigma^2_{Y_h}=1.5$	$\sigma^2_{Y_h}=2.0$	$\sigma^2_{Y_h}=2.5$	$\sigma^2_{Y_h}=3.0$
1.5	58	107	154	202	249	295
2.0	24	42	59	76	93	109
2.5	16	27	37	47	57	67
3.0	13	20	28	35	42	49

8.6 Using the group mean to characterize acceptable exposure

Some investigators have focused on the appropriateness of the group mean exposure level for determining acceptable exposure conditions (Hewett, 1996; Rappaport, 1991b; Rappaport et al., 1988a; Rock, 1982) and on the development of statistical methods for comparing the means of lognormal distributions to OELs (Coenen and Riediger, 1978; Evans and Hawkins, 1988; Galbas, 1979; Hewett, 1997b; Lyles and Kupper, 1996; Rappaport and Selvin, 1987). Strategies for evaluating group mean exposures relative to OELs have been applied to underground mines in the U.S. (Corn, 1985; Corn et al., 1985) and for monitoring long-term exposures to hazardous substances in Germany (Heidermanns et al., 1980; Riediger, 1986).

Various test statistics have been proposed to test whether the group mean exposure $\mu_{X_h} <$ OEL at an α level of significance, notably those due to Rappaport and Selvin (1987), Lyles and Kupper (1996), and Land (1988). Although uniformly the most powerful, the test of Land requires special tables and/or graphs to perform the calculations and often involves various forms of interpolation as well. The other two methods do not require special tables and are simpler to apply. Simulation studies have shown that the method of Lyles and Kupper is more powerful than that of Rappaport and Selvin and performs essentially as well as that of Land with 10 or more observations (Hewett, 1997b; Lyles and Kupper, 1996).

As with the test of exceedance, a test of the group mean exposure requires that all observations be mutually independent, and thus would be restricted to applications under Model (4.1) where a single measurement is obtained from each randomly selected person in a sample (discussed in Chapter 4, Section 4.8). Lyles and Kupper (1996) presented sample sizes required to provide statistical evidence that μ_{X_h} < OEL, with a significance level of $\alpha = 0.05$ and power of 0.80, for situations where the ratio OEL/μ_{X_h} is between 1.5 and 4.0 (reproduced as Table 8.4). The sample sizes indicate that it should be practical to apply the test in situations where μ_{X_h} is less than a fourth to a half of the OEL, except in cases where $\sigma_{Y_h}^2 > 1.0$. For example, using the estimated parameters for Group 2, it appears that slightly more than 7 workers would be required to provide statistical evidence that μ_{X_h} < OEL (OEL = PEL = 3.2 mg/m^3, OEL/$\hat{\mu}_{X_h}$ = 3.8, $\hat{\sigma}_{Y_h}^2$ = 0.50).

Table 8.4 Number of workers (k_h) required for monitoring (one measurement per worker) to demonstrate that the group mean exposure (μ_{X_h}) is less than the OEL with a significance level of $\alpha = 0.05$ and power = 0.80 [from Lyles and Kupper (1996)].

OEL/μ_{X_h}	$\sigma_{Y_h}^2$=0.35	$\sigma_{Y_h}^2$=0.50	$\sigma_{Y_h}^2$=1.0	$\sigma_{Y_h}^2$=2.0	$\sigma_{Y_h}^2$=3.0
1.5	27	38	83	191	322
2.0	12	16	32	71	118
4.0	5	7	11	22	35

8.6.1 Connections to probabilities of exceeding the OEL

Because the group mean μ_{X_h} and variance $\sigma_{X_h}^2$ of the lognormally distributed exposure random variable X_{hij} for Group h are not functionally independent, the levels of exceedance and the probability of overexposure for that group are maximal at a given value of $\frac{OEL}{\mu_{X_h}}$ (Rappaport et al., 1988a). This is shown by substituting $\mu_{X_h} = e^{\left(\mu_{Y_h} + 0.5\sigma_{Y_h}^2\right)}$ into the expressions for $z_{1-\gamma_h}$ and $z_{1-\theta_h}$ given earlier, and solving for $\frac{OEL}{\mu_{X_h}}$; then, we find that

$$\frac{OEL}{\mu_{X_h}} = e^{0.5\left(z_{1-\gamma_h}^2 - (z_{1-\gamma_h} - \sigma_{Y_h})^2\right)}, \text{ and} \qquad (8.8)$$

Probabilities of Exceeding OELs

$$\frac{OEL}{\mu_{X_h}} = e^{0.5\left(z_{1-\theta_h}^2 - (z_{1-\theta_h} - \sigma_{bY_h})^2\right)}. \tag{8.9}$$

From Equations (8.8) and (8.9), we see that the ratio $\frac{OEL}{\mu_{X_h}}$ is maximal when either $z_{1-\gamma_h} = \sigma_{Y_h}$ or $z_{1-\theta_h} = \sigma_{bY_h}$; we will refer to this maximal ratio for Group h as $\left(\frac{OEL}{\mu_{X_h}}\right)_{h,max}$, and the corresponding standard normal variates as $z_{1-\gamma_{h,max}}$ and $z_{1-\theta_{h,max}}$. Note that values of $z_{1-\gamma_{h,max}}$ and $z_{1-\theta_{h,max}}$ are related to the corresponding values of $\left(\frac{OEL}{\mu_{X_h}}\right)_{h,max}$ as follows:

$$z_{1-\gamma_{h,max}} = \sigma_{Y_h} = \sqrt{(2)\ln\left(\frac{OEL}{\mu_{X_h}}\right)_{h,max}}, \text{ and} \tag{8.10}$$

$$z_{1-\theta_{h,max}} = \sigma_{bY_h} = \sqrt{(2)\ln\left(\frac{OEL}{\mu_{X_h}}\right)_{h,max}}. \tag{8.11}$$

The corresponding maximum values of exceedance and probability of overexposure for Group h are designated as $\gamma_{h,max}$ and $\theta_{h,max}$, respectively. Table 8.5 lists values of $\gamma_{h,max}$ and $\theta_{h,max}$ for $2 \leq \left(\frac{OEL}{\mu_{X_h}}\right)_{h,max} \leq 32$, from which we see that when $\left(\frac{OEL}{\mu_{X_h}}\right)_{h,max} = 2$ (so that $z_{1-\gamma_{h,max}} = \sigma_{Y_h} = 1.177$ or $z_{1-\theta_{h,max}} = \sigma_{bY_h} = 1.177$), then $\gamma_{h,max}$ or $\theta_{h,max} = 0.120$. Therefore, if the group mean exposure μ_{X_h} is less than half of the OEL, no more than 12 percent of the exposures from the corresponding lognormal distribution for X_{hij} should exceed the OEL. Likewise, when $\left(\frac{OEL}{\mu_{X_h}}\right)_{h,max} = 4$ (so that $z_{1-\gamma_{h,max}} = \sigma_{Y_h} = 1.665$ or $z_{1-\theta_{h,max}} = \sigma_{bY_h} = 1.665$), then $\gamma_{h,max}$ or $\theta_{h,max} = 0.048$; thus, if the group mean is less than a quarter of the OEL, no more than 5 percent of X_{hij} values should exceed the OEL.

Table 8.5 Maximum values of exceedance ($\gamma_{h,max}$) and probability of overexposure ($\theta_{h,max}$) for given values of $\left(\frac{OEL}{\mu_{X_h}}\right)_{h,max}$.

$\left(\frac{OEL}{\mu_{X_h}}\right)_{h,max}$	$z_{1-\gamma_{h,max}}$ or $z_{1-\theta_{h,max}}$	$\gamma_{h,max}$ or $\theta_{h,max}$
2	1.177	0.120
4	1.665	0.048
8	2.039	0.021
16	2.355	0.009
32	2.633	0.004

8.6.2 Application to STELs

Various implications of Equations (8.10) and (8.11) have been discussed (Esmen, 1992; Rappaport et al., 1988a). One potentially useful application concerns the assessment of exposure levels relative to STELs. Since the mean of a lognormal distribution is independent of the averaging time of measurements (Spear et al., 1986), it is theoretically possible to evaluate the probability that short-term exposures for a typical person in Group h would exceed a STEL based solely upon knowledge of μ_{X_h} (Rappaport et al., 1988a). For example, if it can be demonstrated that μ_{X_h} < STEL/4, then we can infer from Table 8.5 that no more than 5 percent of exposure levels for a typical worker from that group should exceed the STEL during any 15 min (or other) period.

Since μ_{X_h} can be estimated with measurements covering the full work shift, it should be possible to evaluate the frequency of transient excursions of exposure measurements above a STEL *without collecting short-term data*. Rappaport et al. (1988a) explored this notion with continuous exposure data obtained from 41 workers exposed to toluene diisocyanate in 7 factories that manufactured polyurethane foams. In each case that the estimated mean exposure for the ith worker in Group h ($\hat{\mu}_{X_{hi}}$) was less than 0.005 ppm (the TLV-TWA), fewer than 5 percent of 15-min exposures from that worker exceeded the TLV-STEL of 0.02 ppm = 4(TLV-TWA), as predicted by theory.

It is, therefore, implicit in any pair of long-term and short-term exposure limits that, insofar as μ_{X_h} is less than the long-term limit (e.g., TLV-TWA or PEL), the maximum probability that the STEL can be exceeded by a typical worker in that group is

$$\gamma_{STEL_{h,max}} = P\{X_{hij} > STEL\} = 1 - \Phi\left\{2\sqrt{\ln\left(\frac{STEL}{\text{Long-term Exposure Limit}}\right)}\right\}, \quad (8.12)$$

where X_{hij} refers here to the jth short-term exposure measurement on subject i in Group h. Regarding TLVs for benzene (ACGIH, 2007), for example, if μ_{X_h} < TLV-TWA = 0.5 ppm, then the maximum probability that the TLV-STEL = 2.5 ppm would be exceeded for a typical worker in Group h is $\gamma_{STEL_{h,max}} = 1 - \Phi\left\{2\sqrt{\ln\left(\frac{2.5}{0.5}\right)} = 1.79\right\} = 0.037$.

8.7 Regulatory implications

This chapter has considered a complicated milieu in which health professionals interpret the importance of chemical exposure levels relative to OELs. Complications arise because the large variability in air levels, observed both within and between workers, makes it likely that OELs will occasionally be exceeded. While exposure variability puts a premium upon collection of large numbers of measurements (to help explain the sources and magnitudes of exposure levels), the possibilities that OELs will be exceeded implicitly discourages employers from vigorously monitoring their workers' exposures under the current compliance-testing framework. Thus, sample sizes remain small and decisions remain highly capricious; indeed, given likely sample sizes in industrial surveys ($N_h \leq 4$), compliance testing is very likely to allow hazardous conditions to go unnoticed.

Ultimately, our ability to rigorously evaluate exposure levels relative to OELs requires that we accept the fact that there will always be a non-zero probability that an OEL will be exceeded, and thus our goal should be to maintain this probability at an acceptably small value. Regulatory groups that set and enforce exposure limits have been reluctant to embrace this concept, because the notion that exposure levels can allowably exceed an OEL appears to legitimize situations where workers are more highly exposed than desirable. In practice, however, this well-intentioned inflexibility encourages minimal sampling to avoid documenting breaches of the OEL, with the result being that exposure levels remain poorly characterized, some individuals remain highly exposed, and inappropriate control measures are sometimes chosen (as will be seen in Chapter 9). Hopefully, regulators will recognize that their interest in protecting workers' health is not served by the mere appearance of compliance, but rather by encouraging ongoing assessments of exposure levels, using adequate sample sizes to make valid and precise statistical inferences.

Brief coverage was also given to methods about statistical testing of exceedances and the likelihood that group mean exposures exceed the OEL. While both of these approaches can be rigorously evaluated, their applications require independent data [under Model (4.1)] and must, therefore, be limited to single measurements from each worker or to multiple measurements from a single worker. This practical drawback, when coupled with the uncertain connections between exceedances and group mean exposures to individual workers' health risks, makes them less than optimal for evaluating and controlling long-term exposures. Nonetheless, these secondary approaches can have novel applications, such as when considering the juxtaposition of long-term and short-term exposure limits for a given air contaminant.

8.8 This chapter and Chapter 9

In this chapter, we considered two group-level probabilities of exceeding the OEL: the exceedance, which is the probability that a typical group member would experience an air level greater than the OEL on any randomly chosen day (i.e., that $X_{hij} >$ OEL); and, the probability of overexposure, which is the probability that this worker would have a mean exposure level above the OEL (i.e., that $\mu_{X_{hi}} >$ OEL). We showed that the probability of overexposure is more closely connected to the long-term health risks posed by chemical exposures than is the exceedance. In Chapter 9, we will explore the use of the probability of overexposure in a structured strategy for evaluating and controlling long-term exposures in the workplace.

9 INTEGRATING EXPOSURE ASSESSMENT WITH CONTROL

In Chapter 8, we introduced the probability of overexposure for Group h, designated θ_h, as a gold standard for the assessment of long-term exposure levels to toxic chemicals in the workplace. In doing so, we recognized that an observational group can experience substantial variability in exposure levels between persons and that this complicates any assessment of the relationship between exposure and risk of disease. In fact, between-person variability is also a major consideration regarding the control of exposure levels (Rappaport, 1991b). Suppose, for example, that the tasks and personal environments (including locations, types of equipment, and individual work practices) produce long-term-average exposure levels for some workers that are much greater than those for other workers. In such a situation, engineering or administrative controls (which affect workers more-or-less equally) can be of limited effectiveness in reducing exposure levels for highly exposed individuals. Indeed, a more effective strategy would be to identify highly exposed persons so that their personal environments can be modified accordingly to reduce the risk of adverse health outcomes.

In this chapter, we will present a strategy for integrating exposure assessment and control consistent with knowledge of between-person variability. The approach relies upon Models (6.2) and (6.3) to provide the quantification needed to compare exposure levels to an OEL and to evaluate the effects of covariates on control options. We will discuss the salient features of the strategy as well as the statistical methods which are employed, and we will apply the approach to air levels from Groups 1 – 8, as well as to a larger database. Some basic elements of the strategy and its statistical properties have been evaluated in a series of papers (Lyles *et al.*, 1997b; Lyles *et al.*, 1997a; Rappaport *et al.*, 1995a; Weaver *et al.*, 2001), and some features related to θ_h have been examined in industrial applications (Maxim *et al.*, 2000; van Tongeren *et al.*, 2000).

9.1 An integrated strategy

A proposed strategy for assessing and controlling exposure levels is illustrated in Figure 9.1. This strategy provides for grouping workers, for collecting data, and for three stages of decision making. (Issues related to grouping persons and sampling exposures were discussed in Chapter 3). At the first decision stage, Model (6.2) is fit to exposure data for one or more groups, coupled with an examination of residuals and predicted random effects to assess the validity of assumptions and to gauge goodness of fit (as discussed in Chapter 6). If the fit of the model is acceptable, then one proceeds to the second stage, which involves the use of statistical methods for assessing whether exposure levels are acceptable; otherwise, the workers are regrouped using different criteria and Model (6.2) is again fit to the data.

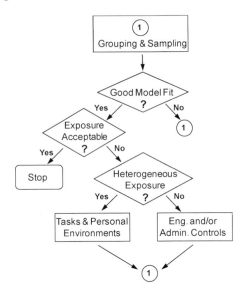

Fig. 9.1 Flow diagram for applying a strategy to assess and control group-specific exposure levels in the workplace.

In testing exposures relative to an OEL, the estimated exposure parameters for Group h are used to determine whether there is sufficient statistical evidence to conclude that the probability of overexposure θ_h is less than an *a priori* acceptable value (designated as A). (We reiterate that this test of overexposure is intended for applications involving long-term exposures to chemicals that produce chronic toxicity and not acute, allergenic, or developmental toxicity). If there is statistical evidence that $\theta_h < A$, then exposure levels are deemed to be acceptable and no additional action is required until the next routine evaluation is performed. If exposure levels are found to be unacceptable, however, the focus then turns to control options for reducing exposure levels.

Optimization of control strategies hinges upon whether exposure levels are *heterogeneous across members of the group*. If so, attention focuses upon tasks and personal environments to identify the individual work elements that contribute to exposure variability across group members. But, if exposure levels are not heterogeneous, then engineering and/or administrative controls are needed to alter the environment for the group as a whole. In either case, the predicted random effects or the values of $_b\hat{R}_{0.95_h}$ (referring to the fold range containing 95% of the estimated individual mean exposure values in Group h) can be used to assess the level of heterogeneity of exposure levels, and Model (6.3) can be applied to identify useful determinants of exposure levels. After selecting a control option, the sampling and statistical evaluation process is repeated until acceptable exposure levels are established.

9.2 Assessing acceptable exposure levels

The statistical test concerning overexposure was motivated by the reasonable assumption (discussed in Chapter 8) that the health risks associated with long-term exposures to chemical agents are related to each individual's cumulative exposure, that is to $CE_{hi}(t) = \mu_{X_{hi}} \times t$, where $CE_{hi}(t)$ is the cumulative exposure for subject i in Group h, and where t represents the number of years of exposure for that subject (Rappaport, 1985; Rappaport, 1991b; Rappaport, 1991a; Rappaport, 1993a; Rappaport et al., 1995a). As discussed in Chapter 8, it is implicit in OELs developed by OSHA that the upper limit of acceptable cumulative exposure would be OEL×45 years. Thus, assuming that a worker can be exposed over a working lifetime of 45 years, then $t = 45$ years and acceptable exposure exists when $\mu_{X_{hi}} <$ OEL. Although this reasoning is strictly valid only when $\mu_{X_{hi}}$ does not vary as a function of t, acceptable risk is still maintained as long as exposure is evaluated regularly and $\mu_{X_{hi}} <$ OEL for each evaluation.

Let A represent the maximum allowable probability that $\mu_{X_{hi}} >$ OEL; acceptable exposure is, therefore, defined by the inequality $\theta_h < A$. In what follows, we will set A equal to 0.10, in keeping with previous applications of the protocol (Lyles et al., 1997b; Lyles et al., 1997a; Rappaport et al., 1995a; Weaver et al., 2001). Although other values of A can be used, sample sizes required to conclude correctly on statistical grounds that exposure levels are acceptable can be very large in cases where A is less than 0.10.

While cumulative exposure should be a valid predictor of long-term health risk, problems can be encountered in extreme situations. Obviously, it would be inappropriate for a worker to be exposed to a lifetime's cumulative exposure, or even a year's cumulative exposure, on one day. For this reason, Rappaport et al. (1995a) suggested that any evidence of individual overexposure (when, for example, the predicted value of an individual worker's mean exposure was greater than the OEL) should be dealt with directly. Again, we emphasize that cumulative exposure should not be used as a measure of risk for acute, allergenic, or reproductive effects.

9.2.1 Testing exposure for a group

In order to test whether or not exposure levels can be deemed to be acceptable for Group h, we consider the following two hypotheses:

H_0: $\theta_h \geq A$ (unacceptable) and H_1: $\theta_h < A$ (acceptable);

thus, rejection of H_0 in favor of H_1 using an appropriate statistical test would lead to a statistical decision of acceptable exposure levels. (Note that rejection of the null hypothesis H_0 in favor of the alternative hypothesis H_1 requires statistical evidence to declare exposure levels acceptable and does not assume acceptable exposure levels as the null state!) A condition equivalent to $\theta_h < A$ is $(\mu_{Y_h} + 0.5\sigma^2_{wY_h} + \sigma_{bY_h} z_{1-A}) < \ln(OEL)$, where z_{1-A} represents the $100(1-A)^{th}$ percentile of the standard normal distribution; e.g., for $A = 0.10$, $z_{1-A} = 1.282$. Thus, by defining the equality

$$R_h = \mu_{Y_h} + 0.5\sigma^2_{wY_h} + \sigma_{bY_h} z_{1-A} - \ln(OEL), \quad (9.1)$$

an equivalent set of hypotheses can be stated as:

H_0: $R_h \geq 0$ (unacceptable) and H_1: $R_h < 0$ (acceptable).

Since μ_{Y_h}, $\sigma^2_{wY_h}$ and $\sigma^2_{bY_h}$ are unknown, Lyles and coworkers (Lyles et al., 1997b; Lyles et al., 1997a; Rappaport et al., 1995a) developed a Wald-type test statistic for testing H_0 versus H_1. Under Model (6.2) with n_{hi} measurements for the ith person in a sample of k_h workers in the hth group, the Wald-type statistic $\hat{W}_h$ for testing H_0 versus H_1 can be written in terms of the estimated values of R_h and its variance, as follows (Weaver et al., 2001):

$$\hat{W}_h = \frac{\hat{R}_h}{\sqrt{\hat{V}[\hat{R}_h]}}, \text{ where} \quad (9.2)$$

$\hat{R}_h = \hat{\mu}_{Y_h} + 0.5\hat{\sigma}^2_{wY_h} + \hat{\sigma}_{bY_h} z_{1-A} - \ln(OEL)$ and $\hat{V}[\hat{R}_h]$ is the Taylor series approximation of the variance of $\hat{R}_h$ given by

$$\hat{V}[\hat{R}_h] \cong \hat{V}[\hat{\mu}_{Y_h}] + \frac{1}{4}\hat{V}[\hat{\sigma}^2_{wY_h}] + \frac{z^2_{1-A}}{4\hat{\sigma}^2_{bY_h}}\hat{V}[\hat{\sigma}^2_{bY_h}] + \frac{z_{1-A}}{2\hat{\sigma}_{bY_h}}\widehat{Cov}[\hat{\sigma}^2_{wY_h}, \hat{\sigma}^2_{bY_h}]; \quad (9.3)$$

both $\hat{R}_h$ and $\hat{V}[\hat{R}_h]$ involve REML estimates in their computation.

Under H_0, the statistic $\hat{W}_h$ has an approximate standard normal distribution for large values of k_h. Thus, H_0 can be rejected in favor of H_1 (and exposure levels declared acceptable) when $\hat{W}_h < z_\alpha$, where z_α is the 100α percentile value of the standard-normal distribution associated with a one-tailed test and an α-level of significance. This Wald-type test has been studied in detail for balanced sets of exposure data (Lyles et al., 1997b). The methodology compares favorably with more complicated test statistics, except for situations in which the value of the ratio $\sigma_{bY_h}^2 / \sigma_{wY_h}^2$ is large. To account for such behavior (likely due to the reduced accuracy of the normal approximation for small sample sizes), the following simple adjustment to the rejection rule is recommended: if $\hat{\sigma}_{bY_h}^2 / \hat{\sigma}_{wY_h}^2 \geq 0.5$, then reject H_0 at the $\alpha/2$ level of significance.

9.2.2 Sample-size requirements

For a specific value of the ratio μ_{X_h} / OEL using the Wald-type test statistic [Equation (9.2)], Rappaport et al. (1995a) provided a relationship to estimate sample sizes required to achieve a given power (1-β) to reject the null hypothesis H_0: $R_h \geq 0$ at significance level α. To perform the sample size calculation, first find the value of θ_h corresponding to the desired value of μ_{X_h} / OEL, namely, $\theta_h = 1 - \Phi\left[\dfrac{\sigma_{bY_h}}{2} - \dfrac{\ln\left(\dfrac{\mu_{X_h}}{\text{OEL}}\right)}{\sigma_{bY_h}}\right]$. Then, find the smallest positive integer k_h such that

$$k_h \geq 1 + \frac{(z_{1-\beta} - z_{1-\alpha})^2 C}{D^2}, \text{ where} \tag{9.4}$$

$$C = \frac{2n_h(n_h-1)\sigma_{bY_h}^2(n_h\sigma_{bY_h}^2 + \sigma_{wY_h}^2) - 2n_h\sigma_{bY_h}\sigma_{wY_h}^4 z_{1-A} + n_h^2\sigma_{bY_h}^2\sigma_{wY_h}^4 + z_{1-A}^2\left[(n_h-1)(n_h\sigma_{bY_h}^2 + \sigma_{wY_h}^2)^2 + \sigma_{wY_h}^4\right]}{2n_h^2(n_h-1)\sigma_{bY_h}^2},$$

assuming n_h exposure measurements are made for each person, and with $D = (z_{1-\theta} - z_{1-A})\sigma_{bY_h}$. This calculation is easily performed using a spreadsheet for several combinations of values of $n_{hi} = n_h$ (starting at $n_h = 2$) and k_h to determine the optimal pair (n_h, k_h) which minimizes the total number of measurements ($N_h = k_h n_h$) needed from Group h. We recommend that k_h be at least 5 (preferably, closer to 10) persons. If estimates of $\sigma_{wY_h}^2$ and $\sigma_{bY_h}^2$ are unavailable, the conservative strategy is to choose a high value of $\sigma_{wY_h}^2$ and a low value of $\sigma_{bY_h}^2$.

Table 9.1 provides sample sizes based upon Equation (9.4) ($N_h = k_h n_h$ for $n_h = 2$ measurements taken from each of k_h persons in the hth group) needed to declare statistically that exposure levels are acceptable (i.e., to reject H_0 in favor of H_1) via the Wald-type test with 80% power and significance level $\alpha =$

0.05.[17] The values of the variance $\sigma^2_{Y_h}$ and the intraclass correlation ρ_h shown in the table are approximately equal to the 25th, 50th and 75th percentiles of the cumulative distributions of $\hat{\sigma}^2_{Y_h}$ and $\hat{\rho}_h$ estimated by Tornero-Velez et al. (1997) from 179 observational groups of workers. When the group mean exposure is at least half the OEL (μ_{X_h}/OEL ≥ 0.5), it is seen that very large sample sizes would be required to reject H_0 with 80% power. However, as the ratio μ_{X_h}/OEL becomes smaller than 0.500, required sample sizes are not excessive, particularly in cases where $\sigma^2_{Y_h}$ ≤ 1, as is generally observed for continuous processes (see Table 5.3). Values of n_h other than two measurements per person result in somewhat different sample sizes for a given power, depending upon the particular values of $\sigma^2_{Y_h}$ and ρ_h. The optimal values of n_h are consistently between 2 and 4 for the combinations of values of $\sigma^2_{Y_h}$ and ρ_h given in Table 9.1.

Table 9.1 Number of measurements ($N_h = k_h n_h$) required for the Wald-type test of overexposure (n_h = 2 measurements per person, A = 0.10, significance level α = 0.05, and power = 0.80).

μ_{X_h}/OEL	$\sigma^2_{Y_h}$=0.50			$\sigma^2_{Y_h}$=1.0			$\sigma^2_{Y_h}$=2.0		
	ρ_h=0.05	0.25	0.50	ρ_h=0.05	0.25	0.50	ρ_h=0.05	0.25	0.50
0.500	178	138	334	444	764	>1000	>1000	>1000	θ_h>A*
0.333	4	28	34	4	74	108	272	246	424
0.250	4	16	16	4	34	42	142	92	118
0.125	4	4	8	4	12	14	4	26	28

*If θ_h>A, then even an infinite number of observations would not allow exposure levels to be declared acceptable.

Assuming that it is desirable to make at least two measurements on each of 5 persons in Group h (N_h = 10), then all workers would typically be monitored in groups containing 5 or possibly fewer workers. The inability to monitor more workers in such a group (i.e., to increase k_h above 5) has obvious power implications, which can be only partially offset by increasing the number of measurements per person. One possible strategy would be to combine several small groups that still share a common factor (e.g., department or building), but do not have the same job. However, this strategy is likely to increase the between-person variability and, therefore, would make it more difficult to statistically declare exposure levels acceptable. Another option would be to randomly sample each person in situations where k_h < 5 and then use the

---

[17] Sample sizes estimated at a given power depend to some extent upon the particular combination of k_h and n_h used. Thus, the sample sizes indicated for n_h = 2 may not be optimal in some situations and smaller values of N_h could be used for a given power when n_h > 2.

alternative test, described below, to determine whether there is statistical evidence that $\mu_{X_{hi}} <$ OEL for each individual worker[18].

9.2.3 Alternative test when the estimated between-person variance component is zero

When a zero estimate of $\sigma^2_{bY_h}$ is obtained, the Wald-type statistic [Equation (9.2)] behaves inappropriately. In such cases, an alternative test is recommended based on concluding that the N_h observations comprise a random sample of data [since there is statistical evidence that $\sigma^2_{bY_h}$ (and hence ρ_h) = 0], and then testing the group mean exposure μ_{X_h} against the OEL (Weaver et al., 2001). Here, we consider the following hypotheses:

$$H'_0: \frac{\mu_{X_h}}{\text{OEL}} \geq \hat{\eta}_h \text{ (unacceptable) vs. } H'_1: \frac{\mu_{X_h}}{\text{OEL}} < \hat{\eta}_h \text{ (acceptable)},$$

where $\hat{\eta}_h = \exp\left(\frac{\hat{\sigma}^2_{bY_h,0.95}}{2} - z_{1-A}\hat{\sigma}_{bY_h,0.95}\right)$ and where $\hat{\sigma}^2_{bY_h,0.95}$ is the estimated 95% upper bound for $\sigma^2_{bY_h}$ (details for calculating $\hat{\sigma}^2_{bY_h,0.95}$ are provided below). Weaver et al. (2001) have shown that H'_0 can be rejected in favor of H'_1 when

$$t = \frac{\hat{\mu}_{Y_h} - \ln(\hat{\eta}_h \text{OEL})}{\hat{\sigma}_{wY_h}/\sqrt{N_h}} \leq t_{(N_h-1),\alpha}(\hat{\delta}),$$ where $t_{(N_h-1),\alpha}(\hat{\delta})$ is the $(100\alpha)^{\text{th}}$ percentile of a noncentral t distribution with (N_h-1) degrees of freedom and estimated noncentrality parameter $\hat{\delta} = -\sqrt{N_h}\left(\frac{\hat{\sigma}_{wY_h,0.95}}{2}\right)$, where $\hat{\sigma}_{wY_h,0.95}$ is the approximate 95% upper confidence bound for σ_{wY_h}, estimated as $\sqrt{\hat{\sigma}^2_{wY_h} + 1.645\widehat{\text{SE}}(\hat{\sigma}^2_{wY_h})}$. [Note that $\widehat{\text{SE}}(\hat{\sigma}^2_{wY_h})$ is the square root of the estimated asymptotic variance of $\hat{\sigma}^2_{wY_h}$ which can be easily obtained from Proc MIXED in SAS].

The estimated parameter $\hat{\sigma}^2_{bY_h,0.95}$, used in the calculation for $\hat{\eta}_h$, can be obtained as $\hat{\sigma}^2_{bY_h,0.95} = \frac{S^*_h U_h Q_h}{c_h(1+Q_h U_h)}$ (Lyles et al., 1997a; Weaver et al., 2001),

where the terms are defined as follows: $S^*_h = (k_h-1)^{-1}\left[\sum_{i=1}^{k_h}\overline{Y}^2_{hi.} - k_h^{-1}\left(\sum_{i=1}^{k_h}\overline{Y}_{hi.}\right)^2\right]$;

$U_h = \frac{S^*_h}{F_{k_h-1,N_h-k_h,0.05}(MSW_h)} - \frac{1}{M_h}$, where $F_{k_h-1,N_h-k_h,0.05}$ is the 5th percentile of the F

---

[18] Note that, in this case, the 'group' consists of one person!

distribution with (k_h-1) d.f. in the numerator and (N_h-k_h) d.f. in the denominator, MSW_h is the within-person mean squared error for Group h [the same as $\hat{\sigma}^2_{wY}$ defined in Section 5.2.3 for a single group] and M_h is the maximum value of n_{hi} in Group h; $Q_h = \dfrac{k_h}{\sum_{i=1}^{k_h} \dfrac{1}{n_{hi}}}$; and $c_h = \dfrac{\chi^2_{k_h-1,0.05}}{k_h-1}$, where $\chi^2_{k_h-1,0.05}$ is the 5th percentile of the chi-square distribution with (k_h-1) d.f.

9.2.4 Testing for overexposure

Table 9.2 summarizes the results of the test of overexposure as applied to the data from Groups 1 – 8, using OELs which were operative at the times measurements were made (the OEL equals the AL for Groups 1 – 3, equals the PEL for Group 4, and equals the TLV-TWA for Groups 5 – 8). Exposure levels were acceptable for Groups 1 and 2, which had estimated group mean air concentrations (values of $\hat{\mu}_{X_h}$) that were 62% and 53% of the respective OELs. Exposure levels for the other 6 groups were all unacceptable, with estimated mean exposures between 52% of the OEL (Group 8) and 9.2 times the OEL (Group 3).

Figure 9.2 illustrates results from application of the Wald-type test of overexposure to 117 occupational groups described by Weaver et al. (2001), again using $A = 0.10$ as the measure of acceptable exposure. Of these 117 groups, 66 had acceptable exposure levels (56% of all groups) and 51 had unacceptable exposure levels (44% of all groups). As shown in Figure 9.2A, groups with acceptable exposure levels typically had a value of $\hat{\mu}_{X_h}$ = 0.10(OEL) ($\hat{\mu}_{X_h}$/OEL: median = 0.10, interquartile range = 0.02 – 0.18, range = 0 – 0.37). Groups with unacceptable exposure levels, on the other hand, typically had an estimated mean exposure level of 0.77(OEL) ($\hat{\mu}_{X_h}$/OEL: median = 0.77, interquartile range = 0.52 – 1.82, range = 0.1 – 17.6). These outcomes were not unduly influenced by sample sizes because the median numbers of measurements were 18 for unacceptable groups and 24 for acceptable groups, and the cumulative distributions of sample sizes were quite similar (Figure 9.2B).

Table 9.2 Results of evaluating exposure levels of Groups 1 – 8 using the Wald-type test statistic, given by Equation (9.2), to test for overexposure when $A = 0.10$ and $\alpha = 0.05$, with parameters estimated under Model (6.2A) for Groups 1-4 and under Model (6.2B) for Groups 5-8.

Group	$\hat{\sigma}^2_{bY_h}$	$\hat{\sigma}^2_{wY_h}$	$\dfrac{\hat{\sigma}^2_{bY_h}}{\hat{\sigma}^2_{wY_h}}$	$\hat{\mu}_{Y_h}$	$\hat{\mu}_{X_h}$ [a]	OEL [a]	$\dfrac{\hat{\mu}_{X_h}}{OEL}$	$\hat{W}_h$ or t [b]	z_α or $t_{(N_h-1),\alpha}(\hat{\delta})$ [c]	Dec. [f]
1	0.023	0.377	0.060	2.73	18.7	30	0.623	-6.58	-1.64	A
2	0.040	0.459	0.087	-0.422	0.841	1.6	0.526	-6.75	-1.64	A
3	2.45	3.07	0.799	-0.069	14.8	1.6	9.25	3.25	-1.96 [d]	U
4	0.283	0.509	0.556	4.62	150	213	0.707	0.930	-1.96 [d]	U
5	0.000	0.362	0.000	2.58	15.8	5.0	3.16	9.65	-3.38 [e]	U
6	0.152	0.362	0.419	1.84	8.14	5.0	1.62	3.64	-1.64	U
7	0.166	0.362	0.458	0.791	2.87	5.0	0.574	-0.638	-1.64	U
8	0.712	0.362	1.97	0.424	2.61	5.0	0.523	0.238	-1.96 [d]	U

[a] Units are µg/m³ for Group 1 and mg/m³ for all other groups.
[b] Test statistic for Wald-type test ($\hat{W}_h$) or alternative test (t).
[c] Critical value for Wald-type test (z_α) or alternative test ($t_{(N_h-1),\alpha}(\hat{\delta})$).
[d] Tested at a significance level of $\alpha/2$ because $\dfrac{\hat{\sigma}^2_{bY_h}}{\hat{\sigma}^2_{wY_h}} > 0.5$.
[e] Alternative test used because $\hat{\sigma}^2_{bY_h} = 0$.
[f] Decision (A = acceptable, U = unacceptable).

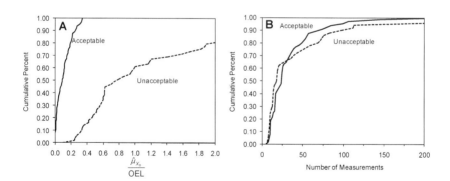

Fig. 9.2. Cumulative distributions of $\hat{\mu}_{X_h}/OEL$ (A) and numbers of measurements (B) for tests of overexposure of 117 groups reported by Weaver et al. (2001). [Based upon the Wald-type test given by Equation (9.2) for $A = 0.10$ and $\alpha = 0.05$; 66 groups had acceptable exposure levels and 51 groups had unacceptable exposure levels].

9.3 Selecting appropriate controls

If a group has unacceptable exposure levels, then concentrations of air contaminants should be reduced by instituting controls. Engineering and administrative controls, which reduce exposures for everyone in the group, will be referred to as *group-level controls*. Other interventions seek to reduce exposures by modifying tasks and/or personal environments (locations, equipment, training, etc.) and thus will be referred to as *individual-level controls*. Since these control options are functionally different, it is important to decide which approach is likely to be more effective for a particular group. The suggested protocol, shown in Figure 9.1, uses the heterogeneity of exposure levels among individual group members to guide this decision; that is, individual-level controls are recommended for groups with heterogeneous exposure levels, and group-level controls are recommended for groups with relatively little heterogeneity in exposure levels.

To determine whether a group has sufficient heterogeneity in subject-specific mean exposure levels to motivate the use of individual-level controls, the protocol employs a simple decision rule based upon the percentage of estimated random-person effects ($\hat{b}_{hi}$) that are significantly different from zero (Rappaport et al., 2003; Rappaport et al., 1999; Weaver et al., 2001). If the number of significant $\hat{b}_{hi}$ values out of k_h such values for Group h is greater than or equal to 10% (i.e., {# of sig. $\hat{b}_{hi}$ }/$k_h \geq 0.10$), then the group is considered heterogeneous. Otherwise, if there are no significant $\hat{b}_{hi}$ values and if $k_h < 10$, then the estimated between-person fold range, $_b\hat{R}_{0.95_h}$, can be used to gauge heterogeneity. From the unweighted least-squares-fitted line in Figure 9.3, which is based upon results from Weaver et al. (2001), a group having $_b\hat{R}_{0.95_h} \geq 10$ is roughly equivalent to one having {# of sig. $\hat{b}_{hi}$ }/$k_h \geq 0.10$. Since the proportion of significant random-person effects [{# of sig. $\hat{b}_{hi}$ }/$k_h \geq 0.10$] is a more rigorous measure of heterogeneity than is $_b\hat{R}_{0.95_h} \geq 10$, we recommend that the latter criterion *only* be used when there are no $\hat{b}_{hi}$ values significantly different from zero and when $k_h < 10$.

We now consider suggested control options for Groups 1 – 8, regardless of the outcomes of the Wald-type tests. The recommended interventions are summarized in Table 9.3, based upon the listed values of {# of sig. $\hat{b}_{hi}$ }/k_h (for all groups except 1 and 5) and $_b\hat{R}_{0.95_h}$ (for Groups 1 and 5). Group-based controls are recommended for Groups 1, 2, 3, 5, and 6, while individual-based interventions are recommended for Groups 4, 7, and 8. (Note that the 9.5% of significant $\hat{b}_{hi}$ values for Group 7 was rounded to 10% for this exercise).

Based upon the information at hand, the above recommendations for controls appear to be reasonable. Referring first to group-level control recommendations, Groups 1, 2, and 3 consisted of operators in the chemical or petrochemical industries where all persons in a given group shared the same sets of tasks and sources of exposure. Likewise, Group 5 was comprised of boiler makers working inside a large vessel at a single work site; here again, tasks and sources of exposure were common to all members. In these four cases, the recommendations for group-based interventions appear to be well-founded. On the other hand, exposures to persons in Group 6 (iron workers) included workers from two sites where work was conducted both indoors and outdoors (see Tables 7.2 and 7.3), and so exposure levels would logically be expected to be quite heterogeneous. Here, the focus upon group-level interventions is not so clear, but may reflect the interaction effect of indoor/outdoor work (IO) with the type of welding process (TW) that was identified in Chapter 7 (see Table 7.5).

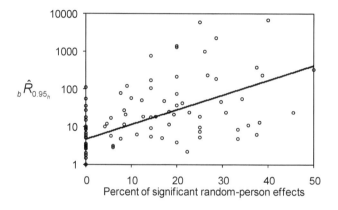

Fig. 9.3. Relationship between the estimated between-worker fold range ($_b\hat{R}_{0.95_h}$) and the percentage of significant random-person effects $\left(\text{i.e.,}\ 100\{\#\ \text{of sig.}\ \hat{b}_{hi}\}/k_h\right)$, for groups with at least 10 sampled workers. Data were compiled from Weaver et al. (2001).

Turning now to the groups where individual-level controls appear to be more appropriate, Group 4 consisted of sprayers and laminators who worked in two different buildings in a boat factory; and, Groups 7 and 8 consisted of pipe fitters and welder fitters who used different types of equipment in welding operations at different sites. In all three of these groups, it is reasonable to expect that exposure levels would vary considerably across persons and so would require some form of intervention at the level of the individual worker. Thus, efforts at controlling exposures to styrene in a boat factory, and to welding fumes among welder fitters and pipe fitters, should initially be directed at identifying the sources of interindividual differences in exposure levels.

Table 9.3 Selection of group-level or individual-level control options for Groups 1 - 8.

Group	$\hat{\mu}_{X_h}/\text{OEL}$	$\hat{\sigma}^2_{bY_h}$	$_b\hat{R}_{0.95_h}$	$\{\#sig.\hat{b}_{hi}\}/k_h$	Control Strategy*
1	0.373	0.023	1.81	0/6	Group
2	0.263	0.040	2.19	0/18	Group
3	4.62	2.45	462	1/24	Group
4	0.707	0.283	8.05	3/19	Individual
5	3.16	0.000	1.00	0/5	Group
6	1.62	0.152	4.61	1/16	Group
7	0.574	0.166	4.94	2/21	Individual
8	0.523	0.712	27.3	4/20	Individual

*Decision rule: if $\frac{100\{\#sig.\hat{b}_{hi}\}}{k_h} \geq 10\%$, then choose an individual-based control; if $\{\#sig.\hat{b}_{hi}\} = 0$, $k_h < 10$, and $_b\hat{R}_{0.95_h} > 10$, then choose an individual-based control option; otherwise, choose a group-based control option.

Referring to the larger database of 117 groups (Weaver et al., 2001), it is informative to determine the likely control paths that would be anticipated more generally under the proposed protocol. Recalling that 51 of these groups had unacceptable exposure levels (using $A = 0.10$), 35 such groups displayed evidence of substantial heterogeneity (69% of all unacceptable groups). For all 117 groups, 65 displayed evidence of substantial heterogeneity (56% of all groups). To the extent that these 117 groups are representative of contemporary industrial exposure groups, this indicates that half to two-thirds of all such groups would require interventions aimed at tasks and/or personal environments, rather than at administrative and/or engineering controls. Since conventional wisdom places a premium on engineering and administrative controls for reducing occupational exposures, this finding suggests that many current efforts at controlling industrial exposures may be misplaced.

9.4 Conclusions

The proposed strategy described in this chapter, which weds a statistical test of overexposure with logical options for controlling potentially harmful exposure levels, has several useful features. First, overexposure can be assessed rigorously in a statistical manner that motivates collection of appropriate and useful exposure data. Second, the connection between long-term health risk and the outcome of the test of overexposure is consistent with assumptions inherent in quantitative risk assessment and with sound toxicological principles. Third, applications of the protocol to over one hundred industrial groups suggest that it is workable with achievable samples of data (the median sample size in these applications was about 20 exposure measurements). Fourth, the finding that more than half of the tested groups showed evidence of significant heterogeneity in exposure levels among group members indicates that the conventional control strategy, specifying group-level interventions exclusively, can be misguided in many cases. And finally, it is worth mentioning that long-term applications of the protocol would ultimately produce comprehensive exposure databases that would be invaluable for use in epidemiologic studies.

9.5 This chapter and Chapter 10

In this chapter, we showed that the parameters estimated under linear mixed Model (6.2) can be used to statistically test the probability of overexposure relative to an OEL and also to select between group-level and individual-level options for controlling exposure levels. This concludes our discussion of exposure assessment for evaluating and controlling workplace hazards. In Chapter 10, we will show how the parameters estimated from some simple linear mixed models can be used to evaluate the effects of exposure measurement errors on estimated exposure-response relationships from epidemiologic studies.

10 EXPOSURE MEASUREMENT ERRORS

10.1 Exposure-response relationships

A major goal of research in occupational and environmental epidemiology is to validly and precisely estimate relationships between levels of exposures to toxic substances and levels of health effects in human populations. If exposure levels are poorly characterized, the estimated exposure-response relationships often underestimate risk for a given exposure (known as *attenuation bias*). Various terms are used to describe the reason for this biasing effect of inaccurate exposure assessment, namely *measurement error* (when exposure is treated as a continuous variable) and *misclassification error* (when exposure is treated as a categorical variable, e.g., as high, medium, and low exposure). In this chapter, we will show how measurement error effects can be related to the underlying random variation in exposure levels within persons, between persons, and between groups. It is assumed that individual exposure measurements randomly vary around the true exposure levels at the times of measurement; that is, there are no systematic errors in measurement of exposure. For example, personal measurements of air contaminants should provide reasonably unbiased estimators of true air levels during the periods of measurement.

The effects of measurement error can be elucidated by comparing a true and an estimated exposure-response relationship. To begin our analysis, we will illustrate the detrimental effects of measurement error using a simple model where a continuous health outcome is linearly related to a continuous exposure variable on the log scale. Examples of continuous health outcomes include quantitative measures of pulmonary function (which can be reduced by exposures to fibrogenic dusts or irritants), concentrations of particular urinary proteins (which reflect kidney damage due to exposures to heavy metals), or frequencies of chromosome aberrations in lymphocytes (which are increased by exposures to genotoxic chemicals). Such a linear relationship is illustrated in Figure 10.1A, where β_1 (> 0) represents the straight-line slope of the true log-scale relationship relating the expected health outcome in a population to the true exposure level. This log-scale linear relationship is able to capture much of the nonlinear behavior observed in exposure-response relationships in the natural scale, as illustrated in Figure 10.1B, where the three straight lines from Figure 10.1A are plotted in the natural scale. When $\beta_1 = 0.5$ in the log scale, the relationship is concave downward (supralinear) in the natural scale, as might be observed when bioactivation of a chemical to a toxic metabolite is saturable. When $\beta_1 = 1.0$ in the log scale, the relationship is linear in the natural scale. And when $\beta_1 = 2.0$ in the log scale, the relationship is concave upward (superlinear) in the natural scale, indicative of a saturable detoxification or repair process. In each case, when the true mean exposure level is zero, it is assumed that the expected health risk is also zero.

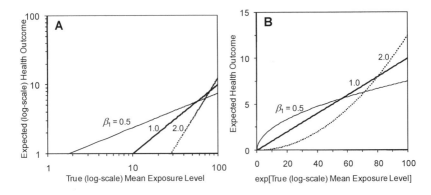

Fig. 10.1 Examples of exposure-response relationships. A: Straight-line relationships between the expected value of the logged continuous health outcome and the true mean logged exposure level, where β_1 is the slope. B: Natural-scale relationships corresponding to the straight lines in A: when $0 < \beta_1 < 1$, the curve is concave downward; when $\beta_1 = 1$, the relationship is linear; and, when $\beta_1 > 1$, the curve is concave upward.

Because the magnitudes and effects of exposure measurement error differ with the type of study design, two types of epidemiologic studies will be considered. First, we will examine an *individual-based study* where exposure levels and health outcomes are measured for all d persons in a random sample. Then, we will consider a *group-based study* where random samples of k persons are measured in each of several groups. In the group-based study, group-specific mean values of both the logged health outcome and logged exposure levels are used to estimate the linear log exposure-log response relationship. For both study designs, we will present formulas that relate the expected values of the estimated regression coefficients, designated $\hat{\beta}_1$ for an individual-based study and $\hat{\beta}_1^*$ for a group-based study, to the true regression coefficient β_1 and the variance components representing exposure variability within-persons, between-persons, and between groups.

In Section 10.6, we will extend the group-based study design to situations where the health outcome is dichotomous (e.g., presence of the disease or not) rather than continuous. The analyses will show under certain assumptions that the same underlying relationship that had been observed between $E(\hat{\beta}_1^*)$ and β_1 for a continuous health outcome also applies to a dichotomous health outcome.

10.2 Individual-based studies

For an individual-based study, we will examine the straight-line relationship (in log-scale) between the true subject-specific mean logged exposure and the expected logged health response in a population. Because we consider a linear log-scale relationship, the true individual exposure level is interpreted as the true mean value of logged exposure levels experienced by that person over time; following our earlier convention, we designate the true mean logged exposure level for the ith person as μ_{Y_i}. Then, the health outcome model for an individual-based study can be written as:

$$R_i = \beta_0 + \beta_1 \mu_{Y_i} + u_i \quad \text{for } i = 1, 2, ..., d \text{ persons;} \quad (10.1)$$

here, R_i represents the natural logarithm of the continuous health outcome; β_0 is the intercept representing the value of $E(R_i|\mu_{Y_i})$ when $\mu_{Y_i} = 0$; β_1 is the true slope relating μ_{Y_i} to $E(R_i|\mu_{Y_i})$; $\mu_{Y_i} = (\mu_Y + b_i)$ is the true unobservable mean of the logged exposure for the ith person [where $b_i \sim N(0, \sigma_{bY}^2)$]; and, u_i is the error term, with $u_i \sim N(0, \sigma_u^2)$. We assume, as under Model (5.1), that the jth logged exposure measurement Y_{ij} for the ith person is modeled as $Y_{ij} = \ln(X_{ij}) = \mu_{Y_i} + e_{ij} = \mu_Y + b_i + e_{ij}$, where $e_{ij} \sim N(0, \sigma_{wY}^2)$. Finally, we assume that all the $\{b_i\}$, $\{e_{ij}\}$, and $\{u_i\}$ are mutually independent random variables. The above assumptions imply that $E(R_i|\mu_{Y_i}) = \beta_0 + \beta_1 \mu_{Y_i}$, $V(R_i|\mu_{Y_i}) = \sigma_u^2$, $E(R_i) = \beta_0 + \beta_1 \mu_Y$, and $V(R_i) = \beta_1^2 \sigma_{bY}^2 + \sigma_u^2$.

In the context of the log-scale relationship depicted in Figure 1A, the x-axis would be μ_{Y_i} and the y-axis would be $E(R_i|\mu_{Y_i})$. The corresponding x- and y-axes in the natural scale (Figure 1B) would be $e^{\mu_{Y_i}}$ (the true geometric mean exposure for the ith person) and $E(e^{R_i}|\mu_{Y_i}) = e^{(\beta_0 + 0.5\sigma_u^2)}(e^{\mu_{Y_i}})^{\beta_1}$, respectively.

10.2.1 Regression analysis

Now, consider a dataset where we have n repeated exposure measurements for each of d randomly selected persons. Since μ_{Y_i} is unobservable, it is reasonable to use $\overline{Y}_i = \frac{1}{n}\sum_{j=1}^{n} Y_{ij}$ as a surrogate measure of μ_{Y_i} and to use the data pairs ($\overline{Y}_i$, R_i), $i = 1, 2, ..., d$, to obtain the *unweighted least-squares estimator*

$\hat{\beta}$ of β_1, where $\hat{\beta} = \dfrac{\sum_{i=1}^{d}(\bar{Y}_i - \bar{\bar{Y}})R_i}{\sum_{i=1}^{d}(\bar{Y}_i - \bar{\bar{Y}})^2}$ and $\bar{\bar{Y}} = \dfrac{1}{d}\sum_{i=1}^{d}\bar{Y}_i$. Since $R_i = \beta_0 + \beta_1 \mu_{Y_i} + u_i$

and $\bar{Y}_i = \mu_{Y_i} + \bar{e}_i = \mu_Y + b_i + \bar{e}_i$, where $\bar{e}_i = \dfrac{1}{n}\sum_{j=1}^{n} e_{ij}$, then the vector $\begin{bmatrix} R_i \\ \bar{Y}_i \end{bmatrix}$ is bivariate normal with mean vector $\begin{bmatrix} \beta_0 + \beta_1 \mu_Y \\ \mu_Y \end{bmatrix}$ and variance-covariance matrix $\begin{bmatrix} \beta_1^2 \sigma_{bY}^2 + \sigma_u^2 & \beta_1 \sigma_{bY}^2 \\ \beta_1 \sigma_{bY}^2 & \sigma_{bY}^2 + \frac{\sigma_{wY}^2}{n} \end{bmatrix}$. So, it follows that

$$E(\hat{\beta}) = \dfrac{\text{Cov}(\bar{Y}_i, R_i)}{V(\bar{Y}_i)} = \dfrac{\beta_1 \sigma_{bY}^2}{\sigma_{bY}^2 + \frac{\sigma_{wY}^2}{n}} = \beta_1 \left(1 + \dfrac{\lambda}{n}\right)^{-1}, \qquad (10.2)$$

where $\lambda = \dfrac{\sigma_{wY}^2}{\sigma_{bY}^2}$ is the *variance ratio*. Equation (10.2) is a well-known result [e.g., see Cochran (1968)] that has been discussed in the context of occupational and environmental epidemiology (Brunekreef et al., 1987; Kromhout et al., 1996; Lin et al., 2005; Rappaport and Kupper, 2004; Tielemans et al., 1998). The standard error of $\hat{\beta}_1$ is given by

$$\text{SE}(\hat{\beta}_1) = \sqrt{\dfrac{n\left[\beta_1^2\left(\dfrac{\sigma_{bY}^2 \sigma_{wY}^2}{n\sigma_{bY}^2 + \sigma_{wY}^2}\right) + \sigma_u^2\right]}{(d-3)(n\sigma_{bY}^2 + \sigma_{wY}^2)}} \qquad (10.3)$$

[see Tielemans et al., (1998)], which indicates that at least four subjects ($d \geq 4$) would be required in an individual-based study to allow the standard error to be estimated using Equation (10.3).

From Equation (10.2), we see that $E(\hat{\beta}_1)$ is attenuated (or suppressed) toward zero by the quantity $\left(1 + \dfrac{\sigma_{wY}^2}{n\sigma_{bY}^2}\right)^{-1}$, referred to as the *reliability coefficient* (Armstrong, 1998). The amount of attenuation increases with σ_{wY}^2 and decreases with both σ_{bY}^2 and n. Since attenuation decreases as σ_{bY}^2 increases, it is wise to select subjects covering the widest possible range of exposures.

10.2.2 Estimating sample sizes

Let $B = E(\hat{\beta}_1)/\beta_1$; then, a measure of relative bias of $\hat{\beta}_1$ as an estimator of β_1 is simply $(B-1)$. For example, if $B = E(\hat{\beta}_1)/\beta_1 = 0.8$, then the relative bias in

$E(\hat{\beta}_1)$ is $(0.8-1) = -0.20$, indicating that $\hat{\beta}_1$ tends, on average, to underestimate β_1 by 20%. Since $B = E(\hat{\beta}_1)/\beta_1 = \left(1+\frac{\lambda}{n}\right)^{-1}$, sample sizes can be estimated from the following relationship for given values of B and λ: $n = \left(\frac{B}{1-B}\right)\lambda$. From Lin et al. (2005), the median value of $\hat{\lambda}$ was 1.1 for 17 occupational studies and was 4.7 for 33 environmental studies. Thus, in order to limit the attenuation bias in $\hat{\beta}_1$ to at most 20% (on average), one would choose $n = \left(\frac{0.8}{1-0.8}\right)(1.1) = 4.4$, meaning at least 5 measurements per person for a typical occupational study, or $\left[\text{since } \left(\frac{0.8}{1-0.8}\right)(4.7) = 18.8\right]$ at least 19 measurements per person for a typical environmental study. Assuming that at least 10 subjects would be sampled in such studies, typical total sample sizes would be at least 50 measurements for an occupational study and at least 190 measurements for an environmental study.

Since sample sizes with many repeated measurements per subject are difficult to achieve in most realistic situations, the above calculations suggest that individual-based studies will generally have non-negligible bias (particularly those conducted in environmental settings), assuming that the simple model structure and attendant assumptions used here are reasonable.

10.3 Group-based studies

10.3.1 Adding a random group effect

Suppose that there are H observational groups ($h = 1, 2, ..., H$) of persons to be investigated for exposure-related health effects. To gain some insight about attenuation bias resulting from studies of this type, Kromhout et al. (1996) considered the following model for $Y_{hij} = \ln(X_{hij})$, where X_{hij} is the jth randomly selected exposure measurement for the ith randomly chosen member of the hth group ($j = 1, 2, ..., n; i = 1, 2, ..., k; h = 1, 2, ..., H$):

$$Y_{hij} = \ln(X_{hij}) = \mu_Y + a_h + b_{hi} + e_{hij} = \mu_{Y_{hi}} + e_{hij} \qquad (10.4)$$

In Model (10.4), $\mu_Y = E(Y_{hij})$ is the overall mean (log-scale) exposure level, $a_h \sim N(0, \sigma^2_{bgY})$ is the random effect for Group h, $b_{hi} \sim N(0, \sigma^2_{bY|h})$ is the random effect for the ith person in Group h (note that the between-subject variance component $\sigma^2_{bY|h}$ can also be regarded as the *within group variance component* reflecting variation across subjects in Group h), and $e_{hij} \sim N(0, \sigma^2_{wY})$ is the random error of the jth exposure measurement for the ith person in the hth group. Note that, in Model (10.4), a_h is the *random effect* of the hth group rather than the *fixed group effect* (μ_{Y_h}) defined in Model (6.2). Thus, there is a *between-group variance component* σ^2_{bgY} associated with Model (10.4). We assume that the $\{a_h\}$, $\{b_{hi}\}$ and $\{e_{hij}\}$ are mutually independent random variables. It then follows that $\mu_{Y_{hi}} = \mu_Y + a_h + b_{hi} \sim N(\mu_Y, \sigma^2_{bgY} + \sigma^2_{bY|h})$, and $Y_{hij} \sim N(\mu_Y, \sigma^2_{bgY} + \sigma^2_{bY|h} + \sigma^2_{wY})$.

Recall that, in Model (6.2), we had represented the H groups as a set $\{\mu_{Y_h}\}$ of fixed effects (rather than as random effects) because our focus was on hazard control, where attention was paid to the assessment and control of exposures to specific occupational groups of interest. Also, some observational groups are uniquely defined and cannot legitimately be regarded as a random sample from an infinite number of possible groups. For example, Groups 5 – 8 were unique because they were defined by the respective trades of particular workers in the U.S. construction industry. However, in epidemiologic studies, it is common to investigate exposures and health outcomes among groups defined more generally by factory, job, city, etc. Here, it is reasonable to assume that a collection of factories, say, represents a random sample of all possible factories that could be chosen for study (Kromhout et al., 1996), and that the particular factories selected are not of unique interest. Thus, Model (10.4) can be more useful in characterizing exposures for epidemiologic studies than a model with fixed effects delineating different groups. However, under Model (10.4), it is important to realize that σ^2_{wY} and $\sigma^2_{bY|h}$ are assumed to be common to all groups. Based upon our experience, this assumption (of homogeneous σ^2_{wY} and $\sigma^2_{bY|h}$ across groups) can be problematic. In fact, following application of Model (6.2) to 117 sets of occupational exposure data, Weaver et al. (2001) showed that it was inappropriate to assume common σ^2_{wY} and σ^2_{bY} in 43 cases (37%), based upon likelihood ratio tests at a significance level of 0.01. This assumption of homogenous variance should be kept in mind when applying Model (10.4) to characterize exposure levels across groups.

Relatively few studies have applied Model (10.4) to estimate variance components across groups. Tables 10.1 and 10.2 summarize results of two such studies which applied Model (10.4) to air contaminants in several factories (Kromhout et al., 1996) and in several U.S. cities (Rappaport and Kupper, 2004). In these two tables, M refers to the total number of persons sampled, while N refers to the total number of measurements obtained from all subjects. The median values of these estimated variance components (listed at the bottom of these tables) will be used in subsequent sample size calculations.

Table 10.1 Variance components estimated under Model (10.4) for levels of air contaminants among workers in several industries [from Kromhout et al., (1996)].

Industry	Agent	M^*	N	$\hat{\sigma}^2_{bgY}$	$\hat{\sigma}^2_{bY|h}$	$\hat{\sigma}^2_{wY}$
Rubber manufacturing	Inhalable dust	231	617	0.18	1.30	0.48
Rubber manufacturing	Solvents	54	111	2.07	1.59	0.63
Animal feed production	Inhalable dust	157	569	0.56	0.84	1.17
Dry cleaning	Perchloroethylene	23	113	0.00	0.88	0.10
Brick manufacturing	Respirable dust	38	150	0.07	0.29	0.32
Reinforced plastics	Styrene	85	258	0.35	0.22	0.46
Petrochemicals	Benzene	418	1949	0.36	0.30	1.56
Bakeries	Flour dust	212	488	0.53	0.42	0.60
Furniture manufacturing	Solvents	16	42	0.55	0.59	0.20
			Median	**0.36**	**0.59**	**0.48**

*Number of persons sampled.

Table 10.2 Variance components estimated under Model (10.4) for levels of volatile organic compounds among the general population in 5 U.S. cities [from Rappaport and Kupper (2004)].

Contaminant	M^*	N	$\hat{\sigma}^2_{bgY}$	$\hat{\sigma}^2_{bY|h}$	$\hat{\sigma}^2_{wY}$
Benzene	421	523	0.21	0.53	0.71
Chloroform	443	553	0.57	0.00	1.26
Ethyl benzene	445	555	0.18	0.20	0.93
Methyl chloroform	447	558	0.07	0.89	1.17
p-Dichlorobenzene	356	398	0.00	1.06	2.00
Perchloroethylene	447	556	0.06	0.07	1.37
Styrene	532	532	0.10	0.07	1.08
Trichloroethylene	444	553	0.29	0.06	2.20
o-Xylene	444	553	0.07	0.00	1.07
		Median	**0.10**	**0.07**	**1.17**

*Number of persons sampled.

10.3.2 Health-outcome model

We will assume the balanced case where there are H groups, each containing k persons with n measurements per person. Let R_{hi} be the natural logarithm of a continuous health response for the ith person in the hth group. The health outcome model we will consider is:

$$R_{hi} = \beta_0 + \beta_1 \mu_{Y_{hi}} + u_{hi}, \tag{10.5}$$

which has the same general structure as Equation (10.1), where $\mu_{Y_{hi}} = \mu_Y + a_h + b_{hi}$ is now the true (but unobservable) mean logged exposure level for subject i in Group h. We assume that $u_{hi} \sim (0, \sigma_u^2)$, and that the $\{u_{hi}\}$ are independent of the $\{a_h\}$, $\{b_{hi}\}$, and $\{e_{hij}\}$. Under Model (10.5), $E(R_{hi}|\mu_{Y_{hi}}) = \beta_0 + \beta_1 \mu_{Y_{hi}}$, $V(R_{hi}|\mu_{Y_{hi}}) = \sigma_u^2$, $E(R_{hi}) = \beta_0 + \beta_1 \mu_Y$, and $(R_{hi}) = \beta_1^2(\sigma_{bgY}^2 + \sigma_{bY|h}^2) + \sigma_u^2$.

10.3.3 Regression analysis

We will use the estimated mean exposure for the hth group, $\overline{Y}_h = \frac{1}{kn}\sum_{i=1}^{k}\sum_{j=1}^{n} Y_{hij} \sim N[\mu_Y, \sigma_{bgY}^2 + (\sigma_{bY|h}^2/k) + (\sigma_{wY}^2/kn)]$, as a surrogate for $\mu_{Y_{hi}}$ (basically assigning each subject in Group h the estimated mean logged exposure level $\overline{Y}_h$ for that group). Then, the Hk pairs ($\overline{Y}_h$, R_{hi}), $h = 1, 2, \ldots, H$ and $i = 1, 2, \ldots, k$, can be used to obtain the unweighted least-squares estimator $\hat{\beta}_1^*$ of β_1 in Model (10.5),

namely $\hat{\beta}_1^* = \dfrac{\sum_{h=1}^{H}\sum_{i=1}^{k}(\overline{Y}_h - \overline{Y})R_{hi}}{\sum_{h=1}^{H}(\overline{Y}_h - \overline{Y})^2} = \dfrac{\sum_{h=1}^{H}(\overline{Y}_h - \overline{Y})\overline{R}_h}{\sum_{h=1}^{H}(\overline{Y}_h - \overline{Y})^2}$, where $\overline{R}_h = \dfrac{1}{k}\sum_{i=1}^{k} R_{hi}$.

Given the stated models and assumptions, the vector $\begin{bmatrix} \overline{R}_h \\ \overline{Y}_h \end{bmatrix}$ is bivariate normal with mean vector $\begin{bmatrix} \beta_0 + \beta_1 \mu_Y \\ \mu_Y \end{bmatrix}$ and variance-covariance matrix

$\begin{bmatrix} \beta_1^2\left(\sigma_{bgY}^2 + \dfrac{\sigma_{bY|h}^2}{k}\right) + \dfrac{\sigma_u^2}{k} & \beta_1\left(\sigma_{bgY}^2 + \dfrac{\sigma_{bY|h}^2}{k}\right) \\ \beta_1\left(\sigma_{bgY}^2 + \dfrac{\sigma_{bY|h}^2}{k}\right) & \sigma_{bgY}^2 + \dfrac{\sigma_{bY|h}^2}{k} + \dfrac{\sigma_{wY}^2}{kn} \end{bmatrix}$. This leads to the result, shown in Kromhout et al. (1996) and Tielemans et al. (1998), that

$$E(\hat{\beta}_1^*) = \frac{\text{Cov}(\overline{Y}_h, \overline{R}_h)}{V(\overline{Y}_h)} = \frac{\beta_1}{1 + \dfrac{\sigma_{wY}^2}{kn\sigma_{bgY}^2 + n\sigma_{bY|h}^2}}, \tag{10.6}$$

which is the measure of expected attenuation in $\hat{\beta}_1^*$ for group-based studies. Also, the standard error of $\hat{\beta}_1^*$ is given by

$$\text{SE}(\hat{\beta}_1^*) = \sqrt{\frac{\frac{\sigma_u^2}{k}\left(\sigma_{bgY}^2 + \frac{\sigma_{bY|h}^2}{k} + \frac{\sigma_{wY}^2}{kn}\right) + \frac{\beta_1^2 \sigma_{wY}^2}{kn}\left(\sigma_{bgY}^2 + \frac{\sigma_{bY|h}^2}{k}\right)}{(H-3)\left(\sigma_{bgY}^2 + \frac{\sigma_{bY|h}^2}{k} + \frac{\sigma_{wY}^2}{kn}\right)^2}} \quad (10.7)$$

[see Tielemans *et al.* (1998)], indicating that at least four groups ($H \geq 4$) would be required in a group-based study so that the standard error of $\hat{\beta}_1^*$ can be estimated using expression (10.7). From Equation (10.6), we see than $E(\hat{\beta}_1^*)$ is attenuated (or suppressed) toward zero by the reliability coefficient $\left(1 + \frac{\sigma_{wY}^2}{kn\sigma_{bgY}^2 + n\sigma_{bY|h}^2}\right)^{-1}$. Attenuation increases directly with σ_{wY}^2 and decreases with $(kn\sigma_{bgY}^2 + n\sigma_{bY|h}^2)$. Attenuation is decreased by increasing σ_{bgY}^2 and/or $\sigma_{bY|h}^2$, as well as k and n, and these quantities are impacted by the study design. In particular, it is important to select groups, and persons within groups, covering the widest possible ranges of exposures. When each group consists of one person ($k = 1$), Equations (10.6) and (10.7) reduce to Equations (10.2) and (10.3), provided earlier for individual-based studies [when $k = 1$, each person represents a group so that σ_{bY}^2 in Equations (10.2) and (10.3) is equal to ($\sigma_{bgY}^2 + \sigma_{bY|h}^2$) in Equations (10.6) and (10.7)].

10.3.4 Estimating sample sizes

To investigate the sample sizes needed to estimate β_1 with no more than a specified level of bias, let $B^* = E(\hat{\beta}_1^*)/\beta_1$, so that the (relative) bias is (B^*-1). Since $\frac{B^*}{1-B^*} = \frac{kn\sigma_{bgY}^2 + n\sigma_{bY|h}^2}{\sigma_{wY}^2}$, then the number of subjects needed to be sampled per group can be derived from the following relationship for given values of B^*, n, σ_{bgY}^2, $\sigma_{bY|h}^2$, and σ_{wY}^2: $k = \left(\frac{B^*}{1-B^*}\right)\left(\frac{\sigma_{wY}^2}{n\sigma_{bgY}^2}\right) - \left(\frac{\sigma_{bY|h}^2}{\sigma_{bgY}^2}\right)$, where the total number of measurements needed per group is equal to kn.

Some sample size calculations employing the above relationship for k are shown in Table 10.3 using median values of variance components from Tables 10.1 and 10.2 (occupational: $\sigma_{bgY}^2 = 0.36$, $\sigma_{bY|h}^2 = 0.59$, $\sigma_{wY}^2 = 0.48$; environmental: $\sigma_{bgY}^2 = 0.10$, $\sigma_{bY|h}^2 = 0.07$, $\sigma_{wY}^2 = 1.17$). Table 10.3 gives the numbers of subjects per group that would be required to limit attenuation bias to at most 20% of β_1 (i.e., $B^* = 0.8$), along with the total number $N = Hkn$ of measurements required assuming $H = 4$ groups. In order to achieve the same level of bias control, the sample size for a typical environmental study (N between about 150 and 200 for the illustrations in Table 10.3) would be roughly 10 times that for a typical occupational study (N between about 15 and 20).

Table 10.3 Number of measurements per group kn (for k subjects per group and n measurements per subject) that would be required to maintain attenuation bias at no more than 20% of the true β_1 value for typical occupational and environmental studies using a group-based study design.

Type of Study	σ^2_{bgY}	$\sigma^2_{bY|h}$	σ^2_{wY}	n	k	kn	N (H=4)
Occupational	0.36	0.59	0.48	1	4	4	16
Occupational	0.36	0.59	0.48	2	2	4	16
Occupational	0.36	0.59	0.48	3	1	3	12
Environmental	0.10	0.07	1.17	1	47	47	188
Environmental	0.10	0.07	1.17	2	23	46	184
Environmental	0.10	0.07	1.17	3	15	45	180
Environmental	0.10	0.07	1.17	4	11	44	176
Environmental	0.10	0.07	1.17	5	9	45	180
Environmental	0.10	0.07	1.17	6	8	48	192
Environmental	0.10	0.07	1.17	7	6	42	168
Environmental	0.10	0.07	1.17	8	6	48	192
Environmental	0.10	0.07	1.17	9	5	45	180
Environmental	0.10	0.07	1.17	10	4	40	160
Environmental	0.10	0.07	1.17	11	4	44	176
Environmental	0.10	0.07	1.17	12	4	48	192
Environmental	0.10	0.07	1.17	13	3	39	156

10.4 Adjusting estimated regression coefficients

Equations (10.2) and (10.6) can be used to adjust estimated slopes for the attenuation biases due to random errors in exposure measurements. That is, the true value of β_1 can be estimated from either $\hat{\beta}_1$ or $\hat{\beta}_1^*$ and the operative estimated reliability coefficient (Armstrong, 1998). Let $\hat{\beta}_{1,adj}$ represent the estimated slope after adjustment for measurement error; that is, $\hat{\beta}_{1,adj} = \hat{\beta}_1 \left(1 + \frac{\hat{\lambda}}{n}\right)$ from Equation (10.2) for an individual-based study (where $\hat{\lambda} = \frac{\hat{\sigma}^2_{wY}}{\hat{\sigma}^2_{bY}}$) and $\hat{\beta}^*_{1,adj} = \hat{\beta}^*_1 \left(1 + \frac{\hat{\sigma}^2_{wY}}{kn\hat{\sigma}^2_{bgY} + n\hat{\sigma}^2_{bY|h}}\right)$ from Equation (10.6) for a group-based study. Armstrong (1998) states that these simple adjustments do not affect the P-values associated with testing whether the estimated slopes are significantly different from zero, and that they can also be applied to compute measurement error-corrected confidence intervals for the true slopes. However, if the true reliability coefficients are estimated using small samples, the precision of such adjustments will undoubtedly suffer.

10.5 Comparing individual-based and group-based studies

Comparing the relationships that define the amounts of attenuation bias in individual-based studies [Equation (10.2)] and group-based studies [Equation (10.3)], the latter design should lead to less attenuation, due to the moderating effect of the group-specific sample size (kn) on σ_{wY}^2. This reduction in bias has been mentioned by several investigators as a motivation for employing group-based studies in epidemiology (Armstrong, 1998; Kromhout et al., 1996; Seixas and Sheppard, 1996; Tielemans et al., 1998). However, since group-based studies employ larger sample sizes than individual-based studies, the advantage of reduced bias should be weighed against the need for larger numbers of measurements.

In order to determine whether the added costs of a group-based study can be justified by the reduction in attenuation bias relative to an individual-based study, it is helpful to consider the ratio of the expected values of the estimators of β_1 for the two types of studies. This ratio, designated $\frac{E(\hat{\beta}_1)}{E(\hat{\beta}_1^*)}$, is derived from Equations (10.2) and (10.6) as

$$\frac{E(\hat{\beta}_1)}{E(\hat{\beta}_1^*)} = \frac{(\sigma_{bgY}^2 + \sigma_{bY|h}^2)(\sigma_{bgY}^2 + \frac{\sigma_{bY|h}^2}{k} + \frac{\sigma_{wY}^2}{kn})}{(\sigma_{bgY}^2 + \sigma_{bY|h}^2 + \frac{\sigma_{wY}^2}{n})(\sigma_{bgY}^2 + \frac{\sigma_{bY|h}^2}{k})}. \tag{10.8}$$

Table 10.4 illustrates calculations, which employ Equation (10.8), to gauge the reduction in attenuation bias that is possible using a group-based study compared to an individual-based study. Again, these calculations use the median values of variance components for occupational and environmental studies identified in Tables 10.2 and 10.3. The results indicate that values of $\frac{E(\hat{\beta}_1)}{E(\hat{\beta}_1^*)}$ are close to one for the typical occupational study as long as k is at least 2. This suggests that the potential reduction in attenuation bias that would be conferred by a group-based occupational study [with a total of at least $4kn$ measurements since $H \geq 4$ to allow the standard error to be estimated using expression (10.7)] relative to an individual-based occupational study [involving a total of at least kn measurements with $k \geq 4$, because at least 4 persons must be sampled to allow estimation of the standard error via expression (10.3)] would probably not be justified by the increased costs involved with collecting at least 4 times as many measurements. On the other hand, the typical environmental study would probably benefit greatly from a group-based design, since this would reduce attenuation bias by at least 50% as long as $k \geq 4$. Note that the metric used in these calculations, i.e., $\frac{E(\hat{\beta}_1)}{E(\hat{\beta}_1^*)}$, is a *relative* measure of the reduction in bias conferred by a group-based design compared to an individual-based design; in fact, both individual-based and group-based studies could suffer from considerable attenuation bias in a particular application.

Table 10.4 Reduction in attenuation bias for a group-based study compared to an individual-based study. (The column labeled $\frac{E(\hat{\beta}_1)}{E(\hat{\beta}_1^*)}$ represents the ratio of the expected straight-line regression coefficient from an individual-based study to that from a group-based study. Values of this ratio that are less than one point to increased attenuation bias in the individual-based study compared to the group-based study.)

Type of Study	σ^2_{bgY}	$\sigma^2_{bY\|h}$	σ^2_{wY}	k	n	$\frac{E(\hat{\beta}_1)}{E(\hat{\beta}_1^*)}$
Occupational	0.36	0.59	0.48	2	2	0.94
Occupational	0.36	0.59	0.48	2	4	0.97
Occupational	0.36	0.59	0.48	2	8	0.98
Occupational	0.36	0.59	0.48	4	2	0.89
Occupational	0.36	0.59	0.48	4	4	0.94
Occupational	0.36	0.59	0.48	4	8	0.97
Occupational	0.36	0.59	0.48	8	2	0.85
Occupational	0.36	0.59	0.48	8	4	0.92
Occupational	0.36	0.59	0.48	8	8	0.96
Environmental	0.10	0.07	1.17	2	2	0.71
Environmental	0.10	0.07	1.17	2	4	0.77
Environmental	0.10	0.07	1.17	2	8	0.83
Environmental	0.10	0.07	1.17	4	2	0.51
Environmental	0.10	0.07	1.17	4	4	0.60
Environmental	0.10	0.07	1.17	4	8	0.70
Environmental	0.10	0.07	1.17	8	2	0.38
Environmental	0.10	0.07	1.17	8	4	0.49
Environmental	0.10	0.07	1.17	8	8	0.63

10.6 Dichotomous health outcomes

We will now extend the measurement-error scenario, described above for group-based studies, to the situation where the health response is dichotomous (e.g., the presence or absence of a particular health outcome). Using the same model for exposure shown in Equation (10.4), namely, $Y_{hij} = \ln(X_{hij}) = \mu_Y + a_h + b_{hi} + e_{hij} = \mu_{Y_{hi}} + e_{hij}$, with the same assumptions about the random effects, we find the following measurement error model:

$$\overline{Y}_h = \overline{\mu}_{Y_h} + \overline{e}_h \text{ for } h = 1, 2, \ldots, H, \tag{10.9}$$

where $\bar{Y}_h = \frac{1}{kn}\sum_{i=1}^{k}\sum_{j=1}^{n} Y_{hij}$, $\bar{\mu}_{Y_h} = \frac{1}{k}\sum_{i=1}^{k} \mu_{Y_{hi}}$, and $\bar{e}_h = \frac{1}{kn}\sum_{i=1}^{k}\sum_{j=1}^{n} e_{hij}$. Now, we define the dichotomous health outcome R_{hi} as equal to 1 if the ith subject in Group h has (or gets) the adverse health outcome and as equal to 0 otherwise. The variable $\bar{R}_h = \frac{1}{k}\sum_{i=1}^{k} R_{hi}$ thus represents the *proportion* of k randomly selected members of Group h that are positive for the health outcome of interest [note that we used $\bar{R}_h$ to represent the group mean of a (logged) continuous health response in Section 10.3.3].

10.6.1 Regression analysis

We define the conditional probability $\pi_{hi} = E(R_{hi} | \mu_{Y_{hi}})$ to be the probability that subject i in Group h (given true mean exposure level $\mu_{Y_{hi}}$) would have (or get) the health outcome of interest. It then follows that $E(\bar{R}_h | \{\mu_{Y_{hi}}\}) = \bar{\pi}_h$, where $\bar{\pi}_h = \frac{1}{k}\sum_{i=1}^{k} \pi_{hi}$. To model $\bar{\pi}_h$, we typically assume a *logistic model* structure, based upon the logit of $\bar{\pi}_h$ given by $\ln\left(\frac{\bar{\pi}_h}{1-\bar{\pi}_h}\right)$. Thus, given the $\{\mu_{Y_{hi}}\}$ and hence $\bar{\mu}_{Y_h}$, the true regression model relating $\bar{\pi}_h$ to $\bar{\mu}_{Y_h}$ is assumed to have the following structure:

$$\text{logit}(\bar{\pi}_h) = \beta_0 + \beta_1 \bar{\mu}_{Y_h}, \qquad (10.10)$$

where $0 \leq \bar{\pi}_h \leq 1$. Now, for $h = 1, 2, \ldots, H$ groups, we assume that $\text{logit}(\bar{\pi}_h) = \ln\left(\frac{\bar{\pi}_h}{1-\bar{\pi}_h}\right) \doteq \ln(\bar{\pi}_h)$; this would be approximately the case if $\bar{\pi}_h$ were 'small' for all h (the so-called 'rare-disease' situation) (Rosner et al., 1989). So, given this approximation, and making the common assumption of non-differential measurement error, then it can be shown (Rosner et al., 1989) that

$$\ln E(\bar{R}_h | \bar{Y}_h) \doteq \beta_0 + \beta_1 E(\bar{\mu}_{Y_h} | \bar{Y}_h) + \frac{\beta_1^2}{2} V(\bar{\mu}_{Y_h} | \bar{Y}_h) = \beta_0^* + \beta_1^* \bar{Y}_h,$$

where $\beta_1^* = \left(\dfrac{\beta_1}{1 + \dfrac{\sigma_{wY}^2}{kn\sigma_{bgY}^2 + n\sigma_{bY|h}^2}}\right)$. Thus, when the data pairs $\{\bar{Y}_h, \bar{R}_h\}_{h=1}^{H}$ are used to fit the model $\ln[E(\bar{R}_h | \bar{Y}_h)] = \beta_0^* + \beta_1^* \bar{Y}_h$, the expected value of the estimated regression coefficient $\hat{\beta}_1^*$ can be related to the true regression coefficient β_1 by the expression $E(\hat{\beta}_1^*) \doteq \dfrac{\beta_1}{1 + \dfrac{\sigma_{wY}^2}{kn\sigma_{bgY}^2 + n\sigma_{bY|h}^2}}$, which is the same as given in Equation

(10.6) for the case involving a continuous health outcome. Therefore, it appears that Equation (10.6) can be used to evaluate the effects of exposure measurement error in epidemiologic studies employing either continuous or dichotomous health outcomes in straight-line exposure-response models.

10.7 Summary

In this chapter we used straight-line models to evaluate the biasing effects of random exposure measurement errors. The exposure models assumed random effects for group, person, and error, and we considered both continuous and dichotomous health outcomes.

Two types of studies were addressed, namely, individual-based studies (where exposure levels and continuous health outcomes are measured for all randomly selected persons) and group-based studies (where exposure levels and continuous or dichotomous health outcomes are measured on randomly selected members of several groups); in the latter study, estimated group means were used to estimate group-specific individual-level exposures for regression analysis. For both study types, relationships were derived for the attenuation bias in estimated straight-line regression slopes (in log scale) as a function of sample sizes and the variance components representing exposure variability within persons, between persons (within groups), and across groups. The results point to several ways to reduce the biasing effects of exposure measurement errors in epidemiologic studies. First, it is important to seek the widest possible ranges of exposures. If the ranges of exposure are modest, then increases in the numbers of subjects and/or measurements per subject will be required. For both types of studies, the relatively large within-subject variability in occupational and environmental exposure levels puts a premium on obtaining replicate exposure measurements on each subject. Overall biases tend to be smaller for group-based studies (due to the larger sample sizes collected in such studies) and thus point to grouping as a means of reducing attenuation bias (when such grouping can be justified despite the increased costs of additional measurements).

When comparing occupational and environmental sources of exposure, it appears that the typical environmental population will have larger within-subject variability, and smaller between-person and between-group variability, than the typical occupational population. This suggests that considerably larger sample sizes are required to investigate health effects arising from environmental exposures than from occupational exposures. It also suggests that group-based designs will be more useful than individual-based designs for investigating exposure-response relationships in the general population, but not necessarily in occupational populations.

Finally, we note that this chapter has focused almost exclusively on considerations of bias (i.e., validity) rather than variability (i.e., precision). Although we think that validity issues should take precedence over precision issues in the design of research studies, we realize that both validity and precision can be considered simultaneously when designing an investigation.

Also, as mentioned earlier, we have considered a relatively simple regression model-measurement error scenario. For treatment of more complicated measurement error situations applicable to investigations of environmental and occupational health, see Lyles and Kupper (1997) and Lyles and Kupper (2000).

10.8 This chapter and Chapter 11

In this chapter, we showed that variance components, estimated from applications of random effects models to occupational and environmental exposure data, can be used to study the effects of attenuation biases in estimated (log-scale) straight-line exposure-response relationships. This concludes our discussion of quantitative exposure assessment based upon the use of environmental measurements. In Chapter 11, we will consider the impact of exposure variability upon levels of chemicals and their products inside the human body. This opens alternative avenues for assessing exposures, based upon biological measurements.

11 EXPOSURE, DOSE, AND DAMAGE

When using measured levels of contaminants to help quantify exposure-response relationships, it is presumed that repeated exposures to such contaminants give rise to a long-term dose of each chemical in the body, and that this dose is predictive of damage to critical molecules, cells, and tissues, and ultimately, is predictive of the risks of disease. The purpose of this chapter is to develop logical linkages among exposure levels, dose, and damage, so that exposure measurements can be connected to the biological processes that determine risks to human health.

11.1 Processes relating exposure and disease

The relationship between long-term exposure to a toxic substance and the risk of disease involves a series of kinetic and dynamic processes that are illustrated in the *exposure-disease model* shown in Figure 11.1 [based upon Lin *et al.* (2005)]. The model, which depicts a population exposed to varying levels of a genotoxic carcinogen, uses cancer as the health outcome. The input is $\{X_{ij}\}$ which, as in previous chapters, represents the set of n discrete exposures [each averaged over Δt hours (h)], received by the ith person in the population. The chemical must first be absorbed from the environment at rate k_{0i} (designated as the *uptake rate* for the ith person), typically through inhalation or ingestion, to produce a burden (mass of contaminant) in the body. In some instances, the substance is intrinsically genotoxic; that is, it is electrophilic and capable of reacting with DNA to produce a mutation. However, most cancer-causing chemicals are *procarcinogens* that must first be metabolically converted to electrophiles. Thus, in Figure 1, $\{P_{ij}\}$ refers to the levels of the *parent compound* (or procarcinogen), while $\{R_{ij}\}$ represents levels of the *reactive electrophile*. The relative amounts of P_{ij} and R_{ij} at any time depend upon competing rates of *passive elimination* of the parent compound (designated k_{1i}, including excretion in breath and urine), of metabolic *bioactivation* of P_{ij} to R_{ij} (designated k_{2i}), of *detoxification* of R_{ij} (designated k_{3i}) via a stable metabolite (M_{ij}) excreted at rate k_{8i}, and to direct reactions of R_{ij} with nucleophiles to produce other products (designated k_{9i}).

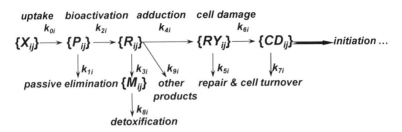

Fig. 11.1 Exposure-disease model using cancer as an example. [Adapted from Lin *et al.* (2005)].

Legend: $\{X_{ij}\}$ is the series of n discrete exposure levels received by the ith person (each averaged over Δt h); $\{P_{ij}\}$ is the series of burdens of the parent compound; $\{R_{ij}\}$ is the series of burdens of a reactive electrophile; $\{M_{ij}\}$ is the series of burdens of a stable metabolite; $\{RY_{ij}\}$ is the series of burdens of a DNA adduct; $\{CD_{ij}\}$ is the series of burdens of damaged cells.

A small fraction of R_{ij} reacts with DNA to produce *adducts* (addition products), some of which are *promutagenic* (because they encourage mismatching of base pairs during the replication of DNA), and hence are capable of inducing tumors. In Figure 11.1, molecular damage is included as the series of promutagenic adduct levels $\{RY_{ij}\}$, representing products of a particular reaction between R_{ij} and nucleophile Y (e.g., a DNA base) which proceeds at rate k_{4i}. Most cells contain repair systems that remove particular DNA adducts and thereby protect the tissue from long-term DNA damage. Also, damaged cells are replaced by normal cells as a consequence of ongoing cell turnover. Thus, the amount of RY_{ij} depends upon the relative rates of *DNA adduction* (i.e., k_{4i}) and *repair* and/or *cell turnover* (designated k_{5i}). Continuing genetic damage over time increases the likelihood that a cell will be initiated, promoted, and ultimately progress to a malignant tumor.

In the exposure-disease model (Figure 11.1), cellular damage is represented by the series $\{CD_{ij}\}$ representing damaged cells, which depends upon the rates of *cell damage* (given by k_{6i}) and *repair* and/or *cell turnover* (at rate k_{7i}). The magnitude of the individual's risk of cancer ultimately depends upon the time integration of CD_{ij} relative to some period of latency, and to that person's susceptibility as determined by genetic, physiologic, metabolic, and lifestyle factors. Note that the rate constants $k_{0i} - k_{9i}$ are assumed to be constant for the ith individual and to be random variables across the population (i.e., to vary from subject to subject).

11.2 Linear kinetics

We will assume that the kinetic processes represented in the exposure-disease model (Figure 11.1) are *linear*. This means that the rate constants are all first-order with respect to removal of the various burdens (P_{ij}, R_{ij}, etc.) from their particular compartments (i.e., k_{1i} and k_{2i} are first-order regarding P_{ij}, k_{3i} and k_{4i}

are first-order regarding R_{ij}, etc.). When levels of exposure are low, as they generally are in the ambient environment, burdens tend to be small relative to those that result in saturation of metabolism, repair, and other saturable processes. Thus, the assumption of linear kinetics is reasonable for populations exposed to environmental contaminants, and for many occupationally exposed groups as well (Hattis, 1998; Olson and Cumming, 1981; Rappaport, 1991b; Rappaport, 1993a). Under linear kinetics, consistent units for all quantities appearing in Figure 11.1 are given in Table 11.1.

Table 11.1 Quantities and their units for the exposure-disease model shown in Figure 11.1.

Quantity	Representing	Units
X_{ij}	Exposure	mg/m³
k_{0i}	Uptake rate	m³/h
P_{ij}	Burden of parent compound	mg
k_{1i}	Elimination rate	1/h
k_{2i}	Bioactivation rate	1/h
R_{ij}	Burden of reactive electrophile	mg
k_{3i}	Detoxification rate	1/h
M_{ij}	Burden of stable metabolite	mg
k_{8i}	Metabolite excretion rate	1/h
k_{4i}	Production rate for RY_{ij}	1/h
RY_{ij}	Burden of DNA adduct (RY)	mg
k_{5i}	Adduct repair rate	1/h
k_{6i}	Cell damage rate	Number of damaged cells per mg RY_{ij}/h
CD_{ij}	Burden of damaged cells	Number of damaged cells
k_{7i}	Cell repair rate	1/h
k_{9i}	Production rate for other adducts	1/h

11.3 The concept of dose

We will define the *dose* for the ith subject as the integral of a particular burden over time for that subject. Thus, D_{P_i} represents the integrated *P*-burden received by the ith person, D_{R_i} the integrated *R*-burden, etc. (Ehrenberg, 1983) Since we consider dose in terms of an integration of burden over time, the units of dose are given as mass-time (e.g., mg-h), or in the case of D_{CD_i}, as the number of damaged cells-time. However, dose can equivalently be presented as the integration (over time) of a tissue concentration within the body, in which case the units would be concentration-time (e.g., mg/l-h). The above definition of dose should not be confused with the *exposure dose*, which refers to the product of external exposure concentration and duration of exposure (with units of mg/m^3-h, say) assuming some constant rate of uptake k_{0i}, or with the *administered dose*, referring to the mass (in mg, say) of a substance administered to a person. Also, the term *biologically effective dose* refers to D_R in the tissue(s) where damage takes place. For example, when considering exposure to benzo(*a*)pyrene [abbreviated as B(*a*)P], a lung carcinogen, we could consider biologically effective dose as the integrated burden of B(*a*)P diolepoxide, a promutagenic metabolite of B(*a*)P, in lung tissue.

11.4 Burden and dose for 'on-off' exposures

We will now investigate the relationship between exposure concentration and D_{P_i} for a simple case, which we will refer to as the 'on-off' exposure scenario.

Under this scenario, exposure levels switch between some constant concentration ('on') and zero ('off') according to a fixed time schedule. The simplest such scenario is depicted at the top of Figure 11.2; at time $t = 0$, the ith subject is exposed to a chemical for exactly 1 h ($t = 1$ h) at constant air concentration $X_i = 1$ mg/m^3. Then, the exposure concentration instantaneously drops to zero ($X_i = 0$ mg/m^3), and the subject is monitored for an additional 1 h (giving a total monitoring time of $t = 2$ h).

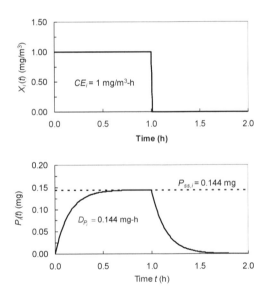

Fig. 11.2 Top: Constant exposure to a chemical for the ith person. Bottom: Burden of the parent compound assuming uptake rate k_{0i} =1 m^3/h and removal rate k_{ri} = 6.93/h; $P_{ss,i}$ represents the value of $P_i(t)$ at steady state.

For the period $0 \leq t < 1$ h at $X_i = 1$ mg/m³, the P-burden at any instant can be described by the following ordinary differential equation:

$$\frac{dP_i(t)}{dt} = k_{0i} X_i - k_{ri} P_i, \quad 0 \leq t < 1 \text{ h; upon integration, we obtain}$$

$$P_i(t) = \frac{k_{0i}}{k_{ri}} X_i (1 - e^{-k_{ri} t}) + P_i(0) e^{-k_{ri} t}, \quad (11.1)$$

where k_{0i} is the rate of uptake and k_{ri} represents the rate of removal of the parent compound by all pathways (regarding Figure 11.1, $k_{ri} = k_{1i} + k_{2i}$) and where $P_i(0)$ is the initial burden (at $t = 0$) for the ith subject. The kinetic relationship depicted in Equation (11.1) is often referred to as a *single-compartment toxicokinetic model*, where the parent compound is assumed to be instantaneously and uniformly distributed throughout the body. Assuming $P_i(0) = 0$, Equation (11.1) reduces to $P_i(t) = \frac{k_{0i}}{k_{ri}} X_i (1 - e^{-k_{ri} t})$, $0 \leq t < 1$ h.

When $t = +\infty$, then $\frac{dP_i(t)}{dt} = 0$, a condition referred to as *steady state*, and $P_i(+\infty) = \frac{k_{0i}}{k_{ri}} X_i = P_{ss,i}$. In our example, $k_{0i} = 1$ m³/h and $k_{ri} = 6.93$/h, corresponding to a half time of $T_{1/2,i} = 0.693/k_{ri} = 0.10$ h; thus, $P_{ss,i} = \frac{1 \text{ m}^3/\text{h}}{6.93/\text{h}} (1 \text{ mg/m}^3) = 0.144$ mg. Therefore, when $t = 1$ h and $k_{ri} = 6.93$/h, $(1 - e^{-k_{ri} t}) = (1 - e^{-0.693(1)}) = 0.999$, so that the subject in our example is essentially at steady state after 1 h of exposure at concentration $X_i = 1$ mg/m³; this is shown at the bottom of Figure 11.2. At $t = 1$ h, X_i drops immediately to 0 mg/m³. So, from Equation (11.1), the P-burden of the subject for the period $1 \leq t < 2$ h is given by $P_i(t) = P_i(1) e^{-k_{ri}(t-1)} \doteq P_{ss,i} e^{-0.693(t-1)}$, $1 \leq t < 2$ h. Thus, at $t = 2$ h, $P_i(2) \doteq 0.144 \, e^{-0.693(2-1)} \doteq 0.0001$ mg $\doteq 0$ mg (see the bottom of Figure 11.2).

The cumulative exposure (CE_i) for our 'on-off' exposure scenario is given by the area under the exposure-time curve, i.e.,

$$CE_i = \int_0^2 X_i(t) dt = \int_0^1 (1) dt + \int_1^2 (0) dt = 1 \frac{\text{mg}}{\text{m}^3}\text{-h}. \text{ And the corresponding}$$

dose of parent compound is given by the area under the P-burden-time curve; i.e.,

$$D_{P_i} = \int_0^2 P_i(t) dt = \int_0^1 \frac{k_{0i}}{k_{ri}} (1)(1 - e^{-k_{ri} t}) dt + \int_1^2 \frac{k_{0i}}{k_{ri}} (1) e^{-k_{ri}(t-1)} dt = \frac{k_{0i}}{k_{ri}} (1) = 0.144 \text{ mg-h}.$$

11.4.1 'On-off' exposure to styrene

Figure 11.3 shows data from Wenker *et al.* (2001), which depicts the mean levels of styrene in the blood of 20 volunteers exposed to either 104 mg/m^3 or 360 mg/m^3 of styrene for one h, followed by two h of non-exposure. The uptake and elimination curves in Figure 11.3 are similar to those shown for the hypothetical situation depicted in Figure 11.2, although the blood levels of styrene clearly did not reach steady state. Wenker *et al.* (2001) reported the estimated mean value for the area under the blood concentration-time curve (*AUC*) as 673 μM-min for the 20 subjects exposed to 104 mg/m^3 of styrene, and as 2276 μM-min for the same 20 subjects exposed to 360 mg/m^3. In this experiment, the estimated *AUC* is analogous to the estimated mean value of D_{P_i} for styrene in the blood of the 20 subjects. Since the *exposure-specific dose*, given by the ratio of the estimated value of the *AUC* to the corresponding exposure concentration, was essentially the same for the two exposure levels (about 6.4 μM-min per mg/m^3), the rates of uptake and elimination of styrene were essentially the same for the two exposure levels of 104 mg/m^3 and 360 mg/m^3.

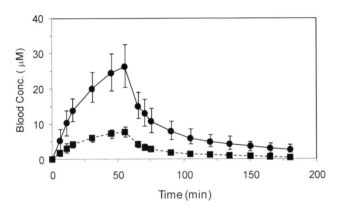

Fig. 11.3. Mean concentrations of styrene in the blood of 20 volunteers exposed to a constant air concentration of styrene for one h [data from Wenker *et al.* (2001)]. Solid curve indicates exposure at 360 mg/m^3; dashed curve indicates exposure at 104 mg/m^3. Error bars represent standard deviations.

Styrene is cleared from the blood by a combination of transport to metabolizing tissues (mainly the liver) and passive elimination via the breath and urine. The decay curve in Figure 11.3, representing elimination of styrene from blood after exposure, displays evidence of the presence of multiple tissue compartments and, therefore, show increasingly slower elimination rates (i.e., multiple values of k_{ri}) as *t* increases. It is common to consider three such compartments for volatile organic compounds such as styrene, namely, a compartment representing highly-perfused tissues (heart, lungs, liver, kidney and spleen - rapid rate of elimination), a compartment representing muscles and skin (intermediate rate of elimination), and a compartment representing the fatty tissues (adipose tissue and bone marrow - slow rate of elimination) [see Clewell et al. (2002) for a review]. From Figure 11.3, it appears that decay of styrene was dominated by the highly-perfused tissues over the first 20 min, by the muscles and skin for the next 40 min, and by the fatty tissues thereafter.

11.4.2 Time to steady state

When considering the P-burden, the time required to reach steady state is a function of the removal rate k_{ri} and is independent of the uptake rate k_{0i}. At a constant level of exposure, the body will achieve about 94% of the steady-state burden of the parent compound in a period equivalent to four half times (i.e., $4T_{1/2,i}$). That is, when $t = 4T_{1/2,i}$ and $P_i(0) = 0$, then, from Equation 11.1, we see that $\dfrac{P_i(4T_{1/2,i})}{P_{ss,i}} = (1-e^{-4T_{1/2,i} k_{ri}}) = (1-e^{-4(0.693/k_{ri})k_{ri}}) = (1-e^{-2.772}) = 0.9375$. Elimination half times, as well as the times to reach approximately 94% of steady state, are presented in Table 11.2 for some common environmental contaminants. These approximate times to steady state cover a remarkable range - from seconds for reactive and irritating chemicals, to years for some heavy metals and lipophilic organic compounds.

Table 11.2 Biological half times for some common contaminants, and approximate times to achieve steady state.

$T_{1/2,i}$ (h)	Contaminant(s)	Time to Steady State*
0 - 1	Reactive chemicals, irritants	< 4 h
1 - 10	Solvent vapors (e.g., benzene and styrene), soluble aerosols in the lung, PAHs in blood	4 – 40 h
10 - 100	Insoluble dusts (rapid), some metals in blood (e.g., nickel), some lipophilic organics (e.g., PCP)	2 – 17 days
100 - 1,000	Insoluble dusts (slow), some metals in blood (e.g., mercury and lead), some lipophilic organics (e.g., PCBs)	2 – 24 weeks
1,000 - 10,000	Some metals in blood (e.g., cadmium), some lipophilic organic compounds (e.g., TCDD)	0.5 – 5 years

Legend: PAHs, polycyclic aromatic hydrocarbons; PCP, pentachlorophenol; PCBs, polychlorinated biphenyls; TCDD, 2,3,7,8-tetrachlorodibenzodioxin.
* Defined as 4 half times, during which the burden would achieve about 94% of the steady state value.

11.5 Random exposure

We will now consider the situation where exposure is described by a random series of air concentrations inhaled by the ith subject on n different days, designated $\{X_{ij}|i\}$ for $j = 1, 2, \ldots, n$ days. As in earlier chapters, we will define $\{X_{ij}|i\}$ as a series of mutually independent, lognormally distributed random air levels with mean value μ_{X_i} and variance $\sigma^2_{X_i}$. We are interested in the corresponding series of P-burdens $\{P_{ij}|i\}$ fixing (or conditional) on the ith subject and in the relationship between $\{P_{ij}|i\}$ and $\{X_{ij}|i\}$. Provided that the daily length of exposure ($\Delta t = 24$ h) is less than the mean residence time of the parent compound (i.e., $1/k_{ri}$), then the approximate burden P_{ij} at the end of the jth day is given by

$$P_{ij} \doteq \frac{k_{0i}}{k_{ri}} X_{ij}(1 - e^{-k_{ri}\Delta t}) + P_{i(j-1)} e^{-k_{ri}\Delta t} , \qquad (11.2)$$

where $P_{i(j-1)}$ represents the P-burden at the end of the preceding $(j-1)^{th}$ day. The first term in Equation (11.2) represents the burden of the parent compound derived from exposure on the current day, while the second term represents the residual burden remaining from all prior days.

Assuming that $\{X_{ij}|i\}$ is stationary[19], then the associated series of P-burdens $\{P_{ij}|i\}$ should also become stationary after sufficient time t ($\gg 1/k_{ri}$) has elapsed. Then, the series $\{P_{ij}|i\}$ can be described in terms of its mean and variance, which we will designate as μ_{P_i} and $\sigma^2_{P_i}$, respectively. Since the expected value of the P-burden should be the same on any day after stationarity is achieved, then $E[P_{ij}] = E[P_{i(j-1)}] = \mu_{P_i}$; and, from Equation (11.2), we find that $\mu_{P_i} \doteq \dfrac{\mu_{X_i} k_{0i}(1 - e^{-k_{ri}\Delta t})}{k_{ri}} + \mu_{P_i} e^{-k_{ri}\Delta t}$, and ultimately that

$$\mu_{P_i} \doteq \frac{k_{0i}}{k_{ri}} \mu_{X_i} . \qquad (11.3)$$

Likewise, $\text{Var}[P_{ij}] = \text{Var}[P_{i(j-1)}] = \sigma^2_{P_i}$; so, from Equation (11.2), we find that $\sigma^2_{P_i} \doteq \sigma^2_{X_i} \left(\dfrac{k_{0i}}{k_{ri}}\right)^2 (1 - e^{-k_{ri}\Delta t})^2 + \sigma^2_{P_i} e^{-2k_{ri}\Delta t}$, and ultimately that

---

[19] As indicated in Chapter 4, the term stationarity implies that the statistical parameters of the underlying process that gives rise to exposure, particularly the mean and variance, do not change over the time period of interest.

$$\sigma_{P_i}^2 \doteq \sigma_{X_i}^2 \left(\frac{k_{0i}}{k_{ri}}\right)^2 \frac{(1-e^{-k_{ri}\Delta t})^2}{(1-e^{-2k_{ri}\Delta t})} = \sigma_{X_i}^2 \left(\frac{k_{0i}}{k_{ri}}\right)^2 \frac{(1-e^{-k_{ri}\Delta t})}{(1+e^{-k_{ri}\Delta t})}, \qquad (11.4)$$

a relationship first reported by Roach (1966).

11.5.1 Dose following random exposure

We will now examine the relationship between a sequence of exposures to a substance, given by $\{X_{ij}|i\}$, and the corresponding average dose of parent compound in the ith subject, given by $\mu_{D_{P_i}}$. Following long-term exposure,

$D_{P_i} = \sum_{j=1}^{n} P_{ij}\Delta t = (n\Delta t)\frac{1}{n}\sum_{j=1}^{n} P_{ij} = t\overline{P_i}$, so that $E(D_{P_i}) = t\mu_{P_i}$; and, from Equation (11.3), $\mu_{P_i} \doteq \frac{k_{0i}}{k_{ri}}\mu_{X_i}$, so that

$$\mu_{D_{P_i}} \doteq t\left(\frac{k_{0i}}{k_{ri}}\mu_{X_i}\right) = \frac{k_{0i}}{k_{ri}}\mu_{CE_i}, \qquad (11.5)$$

where $\mu_{CE_i} = t\mu_{X_i}$ is the expected value of CE_i, representing the cumulative exposure for the ith subject. Therefore, the average long-term dose of the parent compound for the ith subject should, under linear kinetics, be essentially proportional to that subject's cumulative exposure μ_{CE_i}, *regardless of the variance $\sigma_{X_i}^2$ of the exposure series*. This reinforces the notion that cumulative exposure is an inherently meaningful parameter reflecting the true long-term dose received by each individual in an exposed population. While this proportionality between $\mu_{D_{P_i}}$ and μ_{CE_i} is unambiguous under linear kinetics (Hattis, 1998; Olson and Cumming, 1981; Rappaport, 1991b; Rappaport, 1993a), Rappaport *et al.* (2005) empirically showed that a similar proportionality also holds for exposures to volatile organic compounds, even when μ_{X_i} is large enough to saturate metabolism of the parent compound (benzene, perchloroethylene, or acrylonitrile).

11.5.2 Rapid versus slow elimination

Figure 11.4 depicts a hypothetical exposure of a person for 258 consecutive days and the corresponding series of burdens for two chemical substances. The series $\{x_{ij}|i\}$, shown at the top of the figure, was derived from a lognormal distribution with $\mu_{X_i} = 1$ mg/m^3 and $\sigma_{X_i}^2 = 1$ (mg/m^3)^2. The estimated mean exposure $\hat{\mu}_{X_i} = 1.06$ mg/m³ and the estimated standard deviation $\hat{\sigma}_{X_i} = 0.867$, both of which are reasonably close to the true values of 1 mg/m³. Since $t = 258$ d, the estimated cumulative exposure is given by $CE_i = t\hat{\mu}_{X_i} = (258 \text{ d})(1.06 \text{ mg/m}^3) = 274$ (mg/m³)-d.

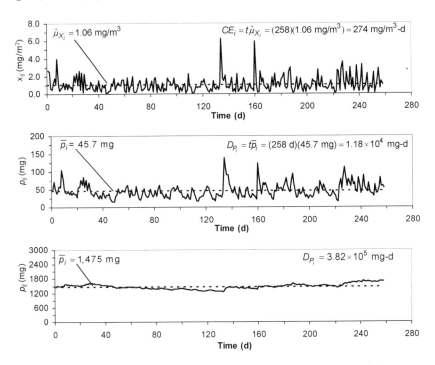

Fig. 11.4 Exposure and dose following random exposure for the ith person. Top: Exposure series $\{x_{ij}|i\}$ for $j = 1, 2, ..., 258$ days with estimated mean value $\hat{\mu}_{X_i}$ and estimated cumulative exposure CE_i. Middle: Series of burdens $\{p_{ij}|i\}$ of the parent compound at the end of each day, assuming uptake rate $k_{0i} = 1$ m³/h and removal rate $k_{ri} = 0.0231$/h ($T_{1/2,i} = 30$ h), with estimated mean P-burden $\bar{p}_i$ and estimated dose D_{P_i}. Bottom: Series of burdens of the parent compound at the end of each day, assuming uptake rate $k_{0i} = 1$ m³/h and removal rate $k_{ri} = 6.93 \times 10^{-4}$/h ($T_{1/2,i} = 1000$ h), with estimated mean P-burden $\bar{p}_i$ and estimated dose D_{P_i}.

The two series of P-burdens shown in Figure 11.4 were predicted from Equation (11.2), assuming $\Delta t = 24$ h and an initial P-burden equal to the true mean value of $\mu_{P_i} = \dfrac{k_{0i}\mu_{X_i}}{k_{ri}}$. Although the uptake rate $k_{0i} = 1$ m³/h was the same for both substances, the middle panel represents the P-burden time series for a relatively rapidly-eliminated chemical with $k_{ri} = 0.0231/h$ ($T_{1/2,i} = 30$ h), while the bottom panel depicts the time series for a relatively slowly-eliminated chemical with $k_{ri} = 6.93 \times 10^{-4}/h$ ($T_{1/2,i} = 1000$ h). Referring to Table 11.1, these half times roughly correspond to elimination of nickel and inorganic lead from the blood, respectively.

For the two substances modeled in Figure 11.4, the estimated mean values of $\{P_{ij}|i\}$, defined as $\overline{P}_i = \frac{1}{n}\sum_{j=1}^{n} P_{ij}$, were $\overline{p}_i = 45.7$ mg for the rapidly-eliminated substance and $\overline{p}_i = 1,475$ mg for the slowly-eliminated substance. This leads to average dose estimates for the hypothetical i^{th} subject of $D_{P_i} = t(\overline{p}_i) = (258 \text{ d})(45.7 \text{ mg}) = 1.18 \times 10^4$ mg-d for the first substance and D_{P_i} equals 3.82×10^5 mg-d for the second substance. These estimated mean doses can be compared with those estimated under Equation (11.5) after 258 days of exposure, which would be $\hat{\mu}_{D_{P_i}} \doteq \frac{k_{0i}}{k_{ri}} CE_i = \frac{1 \text{ m}^3/\text{h}}{0.0231/\text{h}} \times 274$ (mg/m^3)-d $= 1.19 \times 10^4$ mg-d for the first substance and $\hat{\mu}_{D_{P_i}} = 3.95 \times 10^5$ mg-d for the second substance.

11.6 Physiological damping of exposure variability

The two hypothetical series of P-burdens in Figure 11.4 are quite dissimilar in appearance even though they were derived from the same series of exposure levels. The series for the rapidly-eliminated substance (middle of Figure 11.4, $T_{1/2,i} = 30$ h) displays the same overall behavior as that of the exposure series, with prominent fluctuations above and below the mean value on different days. Since this substance is removed from the body rapidly, the residual P-burden from prior days is small; so, from Equation (11.2), $P_{ij} \doteq \frac{k_{0i}}{k_{ri}} X_{ij}(1-e^{-k_{ri}\Delta t}) = c_i X_{ij}$, where $c_i = \frac{k_{0i}}{k_{ri}}(1-e^{-k_{ri}\Delta t})$. Thus, for rapidly-eliminated substances, P_{ij} is proportional to X_{ij} and should share its statistical properties. That is, for rapidly-eliminated substances, since X_{ij} tends to be right-skewed and approximately lognormal, then P_{ij} should also be right-skewed and approximately lognormal.

If the rate of elimination is slow, as when $T_{1/2,i} = 1000$ h (bottom of Figure 11.4), then the time series of P-burdens is very dissimilar to the time series of exposure levels. This is because the current burden is comprised primarily of the residual burden from previous exposures, in which case $P_{ij} \doteq P_{i(j-1)}e^{-k_{ri}\Delta t}$ and so is virtually independent of the current day exposure X_{ij}.

The reduction in variability of a time series of P-burdens relative to a time series of exposure levels was first recognized by Roach (1966), who showed that the ratio of the coefficient of variation (CV) of P_{ij} to that of X_{ij} was a dimensionless function of k_{ri}. That is, from Equations (11.3) and (11.4) we can write:

$$\frac{CV_{P_i}}{CV_{X_i}} = \frac{\sqrt{\sigma_{P_i}^2}\mu_{X_i}}{\sqrt{\sigma_{X_i}^2}\mu_{P_i}} = \left(\frac{1-e^{-k_{ri}\Delta t}}{1+e^{-k_{ri}\Delta t}}\right)^{1/2}. \qquad (11.6)$$

The quantity $\frac{CV_{P_i}}{CV_{X_i}}$ decreases monotonically toward 0 as k_{ri} decreases toward 0 for fixed Δt, and thus this ratio gets smaller for slowly eliminated substances. This phenomenon, which has been termed *physiological damping* (Rappaport, 1985; Rappaport, 1991b; Rappaport and Spear, 1988), has implications for choosing between air samples or biomarkers for exposure assessment (Lin et al., 2005), as will be discussed in Chapter 12.

Referring now to the two hypothetical substances depicted in Figure 11.4, the sample variances of the observed P-burden time series were $s_{P_i}^2 = \frac{1}{n-1}\sum_{j=1}^{n}(p_{ij}-\bar{p}_i)^2 = 385$ mg^2 ($s_{P_i} = 19.6$ mg) for the rapidly-eliminated substance and $s_{P_i}^2 = 8.16\times 10^3$ mg^2 ($s_{P_i} = 90.4$ mg) for the slowly-eliminated substance. Combining these variance estimates with the estimated means given above ($\bar{p}_i = 45.7$ mg for the rapidly eliminated substance and $\bar{p}_i = 1,475$ mg for the slowly eliminated substance), we find values of $\frac{\widehat{CV}_{P_i}}{\widehat{CV}_{X_i}} = \frac{(19.6)(1.06)}{(0.867)(45.7)} = 0.524$ for the rapidly-eliminated substance and $\frac{\widehat{CV}_{P_i}}{\widehat{CV}_{X_i}} = \frac{(90.4)(1.06)}{(0.867)(1,475)} = 0.075$ for the slowly-eliminated substance. Thus, we see that the exposure variability was damped to a much greater extent for the slowly-eliminated substance than for the rapidly-eliminated substance.

Figure 11.5 illustrates the range of $\frac{CV_{P_i}}{CV_{X_i}}$ that would be expected for several toxicants of environmental interest. The figure shows that $\frac{CV_{P_i}}{CV_{X_i}}$ should be close to one for many organic compounds (and their metabolites in blood and urine), and close to zero for some metals (like cadmium) that bind to kidney proteins, for highly lipophilic organic compounds (like TCDD), and for highly insoluble particles (like asbestos) in the lung.

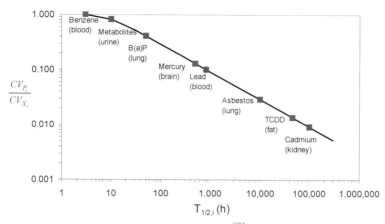

Fig. 11.5 Ratios of coefficients of variation $\frac{CV_{P_i}}{CV_{X_i}}$ for some environmental contaminants of interest [based upon Equation (11.6) using estimates of elimination half times ($T_{1/2,i}$) from literature sources].
Legend: B(a)P, benzo(a)pyrene; TCDD, 2,3,7,8-tetrachlorodibenzodioxin.

11.7 Occupational exposure to mercury

Workers in the chloralkali industry are exposed to inorganic mercury, which is used in large quantities to form cathodes of electrolytic cells producing chlorine and sodium hydroxide. Figure 11.6 depicts time series of air and blood levels of inorganic mercury for a worker in a chloralkali factory. The measurements were reported in summary form by Lindstedt et al. (1979), who kindly provided the original data. During the 54 d of observation of this worker, personal exposure to mercury was measured each work shift and the blood concentration of mercury was measured twice each week. The estimated cumulative exposure and estimated average dose of mercury were $CE_i = t\hat{\mu}_{X_i} = (54 \text{ d})(6.0 \text{ μg/m}^3) = 324 \text{ μg/m}^3\text{-d}$ and

$D_{P_i} = t\bar{p}_i = (54 \text{ d})(90.7 \text{ nM}) = 4.90 \times 10^3 \text{ nM-d}$, respectively. (Note that the estimated mean exposure $\hat{\mu}_{X_i}$ represents the average concentration of mercury in air on all 54 days, assuming an air concentration of zero on non-work days). The series of blood mercury levels displays the physiological damping of exposure variability which is characteristic of slowly eliminated substances. The sample estimates of the CVs for air and blood levels of mercury on the days of blood collection ($n = 14$) were $\widehat{CV}_{X_i} = 1.16$ and $\widehat{CV}_{P_i} = 0.221$, giving the ratio $\frac{\widehat{CV}_{P_i}}{\widehat{CV}_{X_i}} = 0.191$.

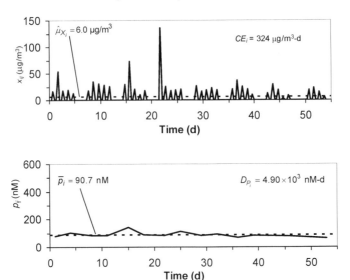

Fig. 11.6 Levels of inorganic mercury in air (top) and blood (bottom) of a chloralkali worker. Data from Lindstedt *et al.* (1979).

11.8 Occupational exposure to styrene

Workers in the fiberglass-reinforced plastics industry are exposed to styrene, which is the major constituent of resin systems used to produce reinforced plastics. When styrene is absorbed in the body, its concentration rapidly equilibrates between the venous blood and the alveolar air. Thus, a portion of the styrene dose is passively cleared from the body by exhalation, and the exhaled breath concentration provides a direct measure of the styrene concentration in mixed venous blood (analogous to the P-burden). Figure 11.7 shows air and breath concentrations of styrene measured repeatedly in 5 workers, performing various jobs in a reinforced plastics factory (manufacturing boats) (Rappaport *et al.*, 1995b). (This is the same study from which the exposure data for Group 4 were obtained; however, the subject numbers in Figure 11.6 do not correspond with those from Group 4). Each subject had pairs of air and breath measurements collected over one year, during 7 surveys conducted at intervals of about 6 weeks. Air measurements were obtained over the full work shift, while breath measurements represent the average values of three randomly-collected samples of mixed-exhaled air collected on that same day.

The pairs of air and breath concentrations of styrene, shown in Figure 11.7, display very similar patterns, and are highly correlated for each of the 4 workers, with Spearman correlation coefficients ranging between 0.679 and 0.892. Such high correlations between daily air and breath levels indicate that styrene is rapidly cleared from the body (as shown previously in Figure 11.2) and that there is little physiological damping of exposure variability in the series of P-burdens (the ratios $\frac{\widehat{CV}_{P_i}}{\widehat{CV}_{X_i}}$ varied between 0.951 and 1.38 for these 5 subjects). Assuming that these workers were exposed to styrene 8 h/d and 5 d/week for a 250 d working year, then their CE_i values ranged from 6.83×10^3 mg/m³-d (subject #2) to 1.07×10^4 mg/m³-d (subject #4) during the year of observation. The corresponding values of D_{P_i} for 250 8-h workdays ranged from 4.07×10^3 μg/m³-d (subject #3) to 9.28×10^3 μg/m³-d (subject #4) (average doses were estimated for each subject using the average breath concentrations over all surveys).

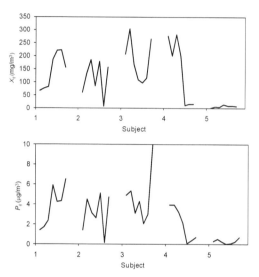

Fig. 11.7 Levels of styrene in air (top) and exhaled breath (bottom) in 5 reinforced plastics workers investigated during 7 surveys about 6 weeks apart. [Data from Rappaport et al. (1995b)].

11.9 Extending the concept of dose

We will now return to the exposure-disease model shown in Figure 11.1 and extend the concept of dose to the time series designated $\{R_{ij}|i\}$, $\{M_{ij}|i\}$, $\{RY_{ij}|i\}$, and $\{CD_{ij}|i\}$. Again assuming linear kinetics and that the duration of exposure is greater than the longest relevant residence time [i.e., $t > \max(1/k_{1i}, 1/k_{3i}, 1/k_{5i}, 1/k_{7i}, 1/k_{8i}$, and $1/k_{9i})$], then the following relationships should apply for large n, where $\mu_{CE_i} = t\mu_{X_i}$:

$$D_{P_i} = \sum_{j=1}^{n} P_{ij}\Delta t \doteq t\mu_{P_i} = \frac{k_{0i}}{(k_{1i}+k_{2i})}\mu_{CE_i} = K_{P_i}\mu_{CE_i}, \qquad (11.7)$$

$$D_{R_i} = \sum_{j=1}^{n} R_{ij}\Delta t \doteq t\mu_{R_i} = \frac{k_{0i}k_{2i}}{(k_{1i}+k_{2i})(k_{3i}+k_{4i}+k_{9i})}\mu_{CE_i} = K_{R_i}\mu_{CE_i}, \qquad (11.8)$$

$$D_{M_i} = \sum_{j=1}^{n} M_{ij} \Delta t \doteq t \mu_{M_i} = \frac{k_{0i} k_{2i}}{(k_{1i} + k_{2i})(k_{3i} + k_{4i} + k_{9i})k_{8i}} \mu_{CE_i} = K_{M_i} \mu_{CE_i}, \quad (11.9)$$

$$D_{RY_i} = \sum_{j=1}^{n} RY_{ij} \Delta t \doteq t \mu_{RY_i} = \frac{k_{0i} k_{2i} k_{4i}}{(k_{1i} + k_{2i})(k_{3i} + k_{4i} + k_{9i})(k_{5i} + k_{6i})} \mu_{CE_i} = K_{RY_i} \mu_{CE_i}, \quad (11.10)$$

$$D_{CD_i} = \sum_{j=1}^{n} CD_{ij} \Delta t \doteq t \mu_{CD_i} = \frac{k_{0i} k_{2i} k_{4i} k_{6i}}{(k_{1i} + k_{2i})(k_{3i} + k_{4i} + k_{9i})(k_{5i} + k_{6i})k_{7i}} \mu_{CE_i} = K_{CD_i} \mu_{CE_i}. \quad (11.11)$$

Equations (11.7) – (11.11) indicate that it is relatively simple to relate the true mean values of the 5 time series in Figure 11.1 under linear kinetics. Note that each dose should be proportional to the mean cumulative exposure (μ_{CE_i}) and that all of the individual kinetic constants can be combined into *accumulation constants*, designated K_{P_i}, K_{R_i}, K_{M_i}, K_{RY_i}, and K_{CD_i} for the ith subject. The products of each of these accumulation constants with μ_{CE_i} represent the integrated burdens of P_i, R_i, M_i, RY_i, and CD_i, respectively.

11.10 This chapter and Chapter 12

In this chapter, we introduced an exposure-disease model to define important functional relationships between a series of exposures received by a particular person and his or her risk of disease, using cancer as an example. Several intermediate processes were identified connecting the exposure time series to health risk, namely, the burdens of the parent compound, of a reactive electrophile, of a stable metabolite, of a DNA adduct, and of damaged cells. Pairs of these processes were linked by rate constants representing uptake, passive elimination, bioactivation, detoxification, adduction, cell damage, and repair or cell turnover. Under linear kinetics, we showed that the true expected dose of a contaminant or its biological endpoints can be expressed as the product of any particular mean burden (e.g., that of the parent compound) and time. This expression is equivalent to the product of the expected cumulative exposure μ_{CE_i} and an accumulation constant representing the particular combination of rate constants describing the relevant toxicokinetics. This connection between expected long-term dose and cumulative exposure emphasizes the importance of cumulative exposure as a measure of individual risk of disease.

When exposure is defined by a random series of air concentrations, the corresponding burden of the parent compound is influenced by the rate of elimination of the contaminant from the body. Burdens of substances that are eliminated rapidly, such as styrene, fluctuate greatly from day to day, in phase with the exposure series. However, slowly-eliminated contaminants, such as mercury, accumulate to large burdens which vary only marginally from day to day. In Chapter 12, we will show how differences in contaminants' rates of elimination can be used, in part, to select between air measurements and biomarkers as surrogates for levels of exposures to toxic substances.

12 BIOMARKERS OF EXPOSURE

As noted in Chapter 1, only about 13% of epidemiologic studies conducted on occupational and environmental populations used any information whatsoever about actual levels of chemical exposures (Armstrong *et al.*, 1992). In addition, those few studies which employed quantitative exposure data relied almost exclusively upon air measurements. In Chapter 1, we also recounted the evolution of air sampling technologies during the 20th century, moving from area measurements to breathing-zone measurements and, finally, to personal measurements. With the current widespread availability of personal air samplers, it is possible to collect large numbers of air measurements and thereby to reduce uncertainties in quantifying levels of airborne exposures for epidemiologic studies.

Biological monitoring has been increasingly viewed as an important alternative to air sampling for assessing occupational and environmental exposures. This technique utilizes biological specimens (especially breath, urine, and blood) to quantify levels of contaminants and/or their products in the body (Rothman *et al.*, 1995). Chemical species or products of cellular damage in these biological specimens are often referred to as *biomarkers*. In Chapter 11, we used biomarkers of mercury and styrene to illustrate phenomena regarding the rates of uptake and elimination of xenobiotic substances in the body. Here, we will consider biomarkers more generally as measures of exposure to toxic chemicals.

12.1 Definitions of biomarkers

As defined by the U.S. National Research Council, biomarkers are *"... indicators of events in biological systems or samples..."*; a biomarker is typically assigned to one of three functional categories (NRC, 1990).

1) *Biomarker of exposure:* *"... an exogenous substance or its metabolite or the product of an interaction between a xenobiotic agent and some target molecule that is measured ... within an organism."*

Examples of biomarkers of exposure include heavy metals in blood or urine, volatile organic compounds in breath or urine, urinary metabolites of organic compounds, and protein adducts of genotoxic substances. In the context of our exposure-disease model (Figure 11.1), most measurements involving the *P*-series, *M*-series, *R*-series, and some *RY*-series (protein adducts) would be biomarkers of exposure.

2) *Biomarker of effect:* "... *a measurable ... alteration within an organism that ... can be recognized as ... health impairment or disease.*"

Examples of biomarkers of effect include promutagenic DNA adducts, chromosome aberrations, decreased lung function, and urinary proteins (reflecting kidney damage). In the context of our exposure-disease model, measurements of agents in the *RY*-series (DNA adducts) and *CD*-series would be biomarkers of effect.

3) *Biomarker of susceptibility:* " ... *an indicator of a ... limitation of an organism's ability to respond to ... a specific xenobiotic substance.*"

Examples of biomarkers of susceptibility include polymorphic forms of metabolism and repair genes. For example, glutathione-*S*-transferases catalyze detoxification reactions between reactive electrophiles and the protective nucleophile glutathione. Polymorphisms of these transferase genes that enhance such reactions should be protective for genetic damage, while those which diminish such reactions would not be protective. In the context of our exposure-disease model, factors that affect many of the rate processes involving *P*-series, *R*-series, *RY*-series and *CD*-series would be regarded as biomarkers of susceptibility and could be treated as potential *effect modifiers* in epidemiologic studies.

12.2 Time scales of biomarkers

As we have seen, some contaminants have much greater residence times ($1/k_i$, where k_i represents one of the removal pathways for the ith person, i.e., k_{1i}, k_{3i}, k_{5i}, k_{7i}, k_{8i}, or k_{9i} in Figure 11.1) and corresponding half times (i.e., $T_{1/2,i} = 0.693/k_i$). Thus, it is important to differentiate among biomarkers according to their residence times (or half times) in the body. We use the following categories, based upon recommendations by Lin *et al.* (2005): short-term biomarkers, where $1/k_i \leq 2$ d; intermediate-term biomarkers, where 2 d $< 1/k_i \leq 2$ months; and long-term biomarkers, where $1/k_i > 2$ months. Based upon this classification scheme, short-term biomarkers reflect exposures during the previous day or two, intermediate-term biomarkers reflect exposures occurring over weeks to months in the past, and long-term biomarkers reflect exposures that have occurred in previous months and years.

The categories of biomarkers defined by residence time have different properties, which make some biomarkers more useful than others for a given application. These properties are summarized in Table 12.1, along with some examples. Short-term biomarkers are best suited for establishing rates of human uptake, elimination, and metabolism of the parent compound. However, due to their short residence times, such biomarkers offer no particular advantages over air samples for dosimetry, health surveillance, and epidemiologic investigations. Intermediate-term and long-term biomarkers, on the other hand, are well suited for dosimetry, health surveillance, and epidemiologic research, and can also be used, in some cases, to define the target dose (e.g., DNA adducts or chromosome aberrations in target tissues). Although biomonitoring has seen only limited applications for elucidating human kinetic relationships, it is becoming increasing clear that the judicious design of studies which employ *both environmental and biological monitoring* can provide valuable information about kinetic processes in human populations [for example, see Johnson *et al.* (2007) and Taylor *et al.* (in press)].

Table 12.1 Properties of biomarkers classified by residence time ($1/k_i$).

Type of Biomarker	Residence Time	Damping	Examples	Best Applications
Short-term	$1/k_i \leq 2$ d	None to moderate	Reactive chemicals in blood; solvents in blood, breath, and urine; most urinary metabolites	Uptake, distribution, elimination and metabolism of parent compound and reactive electrophile
Intermed.-term	2 d $<1/k_i \leq 2$ mo.	Moderate to great	Nickel, cobalt, mercury, and lead in blood; PCP in blood; DNA adducts; PCBs in blood and fat; albumin adducts in blood	Dosimetry, epidemiology, and health surveillance
Long-term	2 mo.$< 1/k_i$	Very great	Cadmium and mercury in urine; TCDD in blood; hemoglobin adducts; lymphocytic SCEs and chromosome aberrations	Dosimetry, epidemiology, and health surveillance

Legend: PCP, pentachlorophenol; PCBs, polychlorinated biphenyls; TCDD, 2,3,7,8-tetrachlorodibenzodioxin; SCEs, sister-chromatid exchanges.

12.3 Choosing between air measurements and biomarkers

12.3.1 Intrinsic advantages

Table 12.2 summarizes the intrinsic advantages and disadvantages of air samples and biomarkers, as surrogate measures of true individual exposure levels in epidemiologic studies. Since biomarkers reside closer to the disease endpoint (see Figure 11.1), they are theoretically more relevant measures of exposure levels for subject *i* than are the observed air levels $\{X_{ij}|i\}$. As shown in Chapter 11, levels of intermediate-term and long-term biomarkers also tend to be less variable than daily air concentrations (due to physiological damping) and, therefore, can provide more precise measures of individual mean exposures. It is also true that biological monitoring accounts for all routes of exposure (i.e., inhalation, ingestion, and dermal absorption), and thereby better reflects the total exposure received by an individual. Biomarkers can also reflect unexpected or accidental exposures that could easily escape detection by air sampling. Finally, biomarkers provide information about interindividual differences in uptake, elimination, metabolism, and repair due to exposures to toxic substances.

Table 12.2 Relative advantages of air measurements and biomarkers as measures of exposures to chemicals.

Measure of Exposure	Relative Advantages
Biomarkers of exposure	Closer to disease endpoints
	Can damp exposure variability
	Reflect all routes and sources of exposure
	Unaffected by personal protection
	Reflect unexpected or accidental exposures
	Provide information about interindividual differences in uptake, etc.
Air measurements	Simpler and cheaper (permit larger sample sizes)
	Less invasive and more acceptable to subjects
	Assays more precise and specific
	Easily related to exposure limits and controls
	Relatively little autocorrelation from day to day

On the other hand, air sampling is generally simpler and less expensive than biomonitoring and can, therefore, provide larger sample sizes for a given study at a given cost. Air measurements also tend to be less invasive and more acceptable to subjects than biomarker measurements. Assays for air measurements are generally more precise and more specific than those for biomarkers due to the much simpler analytical matrix (e.g., air versus blood or urine). Also, the relationship between measurements and the source(s) of exposure is generally clear for air measurements but not necessarily for biomarkers. Indeed, biomarkers representing the RY and CD series are notoriously nonspecific; e.g., N^2-ethenoguanine, a promutagenic DNA adduct, can be produced either by exposure to exogenous ethylene or ethylene oxide, or by endogenous processes (Albertini *et al.*, 2003), and chromosome aberrations can arise from a plethora of chemical agents as well as from ionizing radiation and reactive oxygen species produced endogenously. Because of their connections to sources of exposure, air measurements are much more easily related to OELs and interventions than are levels of biomarkers (which reflect both exogenous sources, in and out of the workplace, and endogenous sources). Although air measurements tend to display little autocorrelation from day to day, biomarker series can be highly autocorrelated when the sampling interval is shorter than the residence time; this adds complexity to experimental design and statistical analyses involving biomarker measurements (Droz *et al.*, 1991; Lin *et al.*, 2005).

12.3.2 Biasing effects of surrogate exposure measures

In Chapter 10, we showed that within-person variability in air concentrations can give rise to measurement error effects that often attenuate estimated exposure-response relationships. Since biomarker levels also vary within persons, the same sort of attenuation bias would occur if biomarkers were used as surrogates for exposure. In addition, the magnitudes of such biasing effects can differ among a given set of candidate biomarkers (e.g., P-, M-, or RY-based biomarkers) for the same chemical exposures. Thus, it is an open question whether measurements of a particular biomarker would be less biasing than air measurements for an epidemiologic study, or even which of several candidate biomarkers would be the least biasing surrogate measure of exposure.

The notion that the biasing potentials of air and biological measurements might be used to guide the selection of a particular exposure surrogate for an epidemiologic study was first proposed by Rappaport *et al.* (1995b) who used data from a single investigation to illustrate the idea. In a later large study of the biasing effects of air samples and biomarkers, Lin *et al.* (2005) compiled a database of paired air and biological measurements (12,077 total measurements) from 43 longitudinal studies of occupational and environmental populations. In what follows, we will summarize the important results from Lin *et al.* (2005) and then make suggestions regarding the use of air measurements or biomarkers for epidemiologic studies.

12.3.3 Estimated variance components

Lin *et al.* (2005) analyzed air levels and biomarkers from studies covering numerous pollutants (mainly metals and volatile organic compounds) in both environmental and occupational settings. The numbers of the various series of biomarkers examined decreased in the following order (see Figure 11.1): P (21), M (12), RY (7), CD (3), and R (2). For some contaminants, more than one biomarker was measured for a given set of air measurements; and, in a few studies, only biomarker measurements were available (77% of the datasets contained paired air and biomarker measurements).

After compiling their database, Lin *et al.* (2005) estimated within-person and between-person variance components under Model (6.3) with a fixed effect for time, where time was variously considered as a linear trend, as a weekday effect, or as a seasonal effect. Significant time effects were found in approximately one-third (18 of 50) of the air monitoring data sets and in approximately half (36 of 77) of the biomarker data sets. Seasonal effects were most commonly observed (43 datasets), followed by weekday effects (7 datasets), and then linear trends (4 datasets). Compound symmetry was found to be appropriate to model the variance-covariance matrices for all datasets of air measurements and for all but 4 biomarker datasets. An exponential covariance matrix provided a better fit for the latter 4 datasets, all involving intermediate-term or long-term biomarkers (DDE and trans-nonachlor in blood, and inorganic lead and δ-aminolevulinic acid in urine).

The cumulative distributions of the estimated between-person and within-person variance components are shown in Figure 12.1 in terms of the corresponding fold ranges (i.e., $_b\hat{R}_{0.95}$ and $_w\hat{R}_{0.95}$, respectively) for air measurements and biomarkers. The distributions of $_b\hat{R}_{0.95}$ were very similar for both types of measurements, with median $_b\hat{R}_{0.95}$ values between about 7-fold and 8-fold (left part of Figure 12.1). However, the cumulative distributions of $_w\hat{R}_{0.95}$ were significantly different (P-value < 0.01) for the two types of surrogates, with biomarker measurements typically being much less variable (median $_w\hat{R}_{0.95} = 17.4$) than air measurements (median $_w\hat{R}_{0.95} = 48.9$) (right part of Figure 12.1). Since about half of the biomarker data sets involved intermediate-term and long-term biomarkers, this reduction in within-person variation probably reflects physiological damping of exposure variability (discussed in Chapter 11). As further evidence supporting this damping effect, when the biomarker data were stratified by residence time, median values of $_w\hat{R}_{0.95}$ decreased in the order: short-term (median = 44.6) > intermediate-term (median = 3.7) > long-term (median = 3.3).

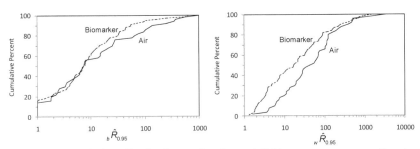

Fig. 12.1 Cumulative distributions of estimated fold ranges corresponding to between-person (left) and within-person (right) variance components for studies in which both air measurements and biomarkers were measured on the same subjects. Data from Lin et al. (2005).

12.3.4 Estimated variance ratios

In Chapter 10, we introduced a simple health-outcome model [Equation (10.1)] which related the (logged) continuous measure of health risk R_i for the ith person to the true *unobservable* mean (logged) exposure (μ_{Y_i}) as follows:

$R_i = \beta_0 + \beta_1 \mu_{Y_i} + u_i$, where β_1 is the true straight-line regression coefficient for exposure. One analysis strategy for an individual-based study of health effects would be to estimate β_1 by substituting a surrogate for μ_{Y_i} in Equation (10.1) for each subject in the sample. In Chapter 10, we considered $\overline{Y}_i = \frac{1}{n}\sum_{j=1}^{n} Y_{ij}$ as this surrogate, where Y_{ij} is the jth (logged) air measurement for the ith person ($j = 1, 2, \ldots, n$). We could just as easily use observed biomarker levels as surrogates for μ_{Y_i}, in which case Y_{ij} would represent the jth of n (logged) biomarker measurements for the ith person. In either case, potential bias in the estimation of the true slope (β_1) of the health-outcome Model (10.1) would be related to the variance ratio $\lambda = \sigma_{wY}^2 / \sigma_{bY}^2$ according to Equation (10.2), where it is seen that bias decreases as λ decreases. In their analyses, Lin et al. (2005) considered potential biasing effects of air measurements and biomarkers by examining the estimated variance ratio $\hat{\lambda} = \hat{\sigma}_{wY}^2 / \hat{\sigma}_{bY}^2$. They found that values of $\hat{\lambda}$ for biomarker measurements (median = 1.04) were significantly smaller than those for air measurements (median = 2.40) (*P*-value = 0.02), and they concluded that biomarkers typically provide less-biasing estimates of the true slope (β_1) than do air measurements.

As shown in Figure 12.2, when the data were stratified by the residence time of the biomarker, median values of the estimated variance ratios decreased in the order short-term median > intermediate-term median > long-term median, and a significant difference was detected between short-term and long-term medians (*P*-value < 0.05). This again points to the effect of physiological damping of exposure variability for intermediate-term and long-term biomarkers.

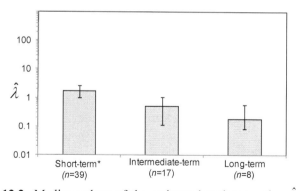

Fig. 12.2 Median values of the estimated variance ratio, $\hat{\lambda}$, for biomarkers classified by residence time. Error bars represent interquartile ranges. Data from Lin et al. (2005).
* Indicates a significant difference between the long-term and short-term median values (*P*-value < 0.05).

12.3.5 Estimated lambda ratios

Finally, Lin et al. (2005) estimated lambda ratios ($\hat{\lambda}_{bio} / \hat{\lambda}_{air}$) for the 54 datasets having sets of both air and biomarker measurements. Almost two-thirds of these datasets (62%) had estimated lambda ratios of less than one (median estimated lambda ratio = 0.46), again providing evidence that biomarkers provide less biasing surrogate measures of exposure than do air measurements. The median and interquartile ranges of estimated lambda ratios were also compared by type of agent, as shown in Figure 12.3. Here, the median estimated lambda ratio was significantly less than one for metal exposures, significantly greater than one for pesticide exposures, and not significantly different from one for organic compounds. While the reduced estimated lambda ratio for metal exposures likely resulted from physiological damping of exposure variability by the slowly-eliminated metals like lead and mercury, the increased estimated lambda ratio for pesticide exposures could indicate that subjects in those studies had absorbed pesticides from dietary and/or dermal sources as well as from inhalation. However, since this finding was based on only 4 pesticide datasets, additional data will be required before making any definitive statements.

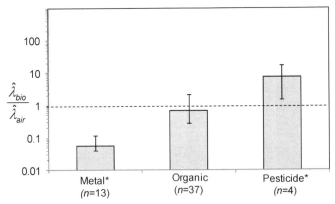

Fig. 12.3 Median values of the estimated lambda ratio, $\hat{\lambda}_{bio} / \hat{\lambda}_{air}$, for biomarkers classified by type of contaminant. Error bars represent interquartile ranges. Data from Lin et al. (2005).
* Indicates a significant difference from one (P-value < 0.05).

Table 12.3 lists the estimated values of lambda ratios compiled by Lin et al. (2005), in order of increasing $\hat{\lambda}_{bio} / \hat{\lambda}_{air}$. Median values of $\hat{\lambda}_{bio} / \hat{\lambda}_{air}$ ranged from 0.031 for lead in blood to 46.0 for a DNA adduct of styrene oxide. The estimated within-person variance components of the 7 biomarkers with the smallest estimated lambda ratios were all considerably smaller than those of the corresponding air measurements for the same subjects, suggesting that day-to-day fluctuations in levels of these 7 biomarkers were damped relative to those of the corresponding air concentrations (data not shown). Indeed, 5 of the 7 biomarkers with the lowest median values of estimated lambda ratios ($0.031 \leq \hat{\lambda}_{bio} / \hat{\lambda}_{air} \leq 0.112$) were related to measurements of metals (lead, mercury, and chromium) or their products in blood or urine, all of which are eliminated very slowly. Among those biomarkers with the 7 lowest estimated lambda ratios, the other two were perchloroethylene in breath (median $\hat{\lambda}_{bio} / \hat{\lambda}_{air} = 0.096$) and a hemoglobin adduct of ethylene oxide in red blood cells ($\hat{\lambda}_{bio} / \hat{\lambda}_{air} = 0.107$). In both of these cases, the damping of exposure variability within subjects is also mechanistically reasonable. Hemoglobin adducts are removed with the turnover of red blood cells as they reach an age of about 120 days in humans; thus, the mean residence time of human hemoglobin adducts should be about 60 days (Tornqvist et al., 2002). Likewise, perchloroethylene is eliminated more slowly than most other volatile organic compounds, due to its high lipophilicity and limited metabolism in humans (Rappaport et al., 2005).

Turning now to the 7 biomarkers having the highest estimated lambda ratios in Table 12.3 ($5.73 \leq \hat{\lambda}_{bio} / \hat{\lambda}_{air} \leq 46.0$), inspection of the estimated variance components paints a different picture. For all of these 7 biomarkers, the estimated within-person variance components were larger for the biological measurements than for the corresponding air measurements, sometimes by large margins (data not shown). For example, an unidentified DNA adduct of styrene oxide had a value of $\hat{\sigma}_{wY}^2 = 2.51$ compared to a value of $\hat{\sigma}_{wY}^2 = 0.131$ for measurements of styrene oxide in the air at a reinforced plastics factory. This huge difference in estimated values of the within-person variance component primarily reflects the extreme imprecision of the ^{32}P-postlabeling assay that was used to measure the DNA adducts in that study (Horvath et al., 1994). Another large increase in $\hat{\sigma}_{wY}^2$, from 0.725 to 1.24, was observed for free styrene glycol in the blood relative to air levels of styrene for the same reinforced plastics workers. This also reflects imprecision in the assay used to measure styrene glycol and the fact that conjugates of styrene glycol, that likely reflected much of the styrene dose, were not included in the analyses. Increases in values of $\hat{\sigma}_{wY}^2$ between pairs of air and biological measurements were not so great for the other 5 biomarkers listed at the bottom of Table 12.3, but were countered by large decreases in the corresponding values of $\hat{\sigma}_{bY}^2$ for biomarker levels relative to air levels in each case (data not shown). This reduction in $\hat{\sigma}_{bY}^2$ for biomarker levels compared to the corresponding air levels was likely caused by dietary and/or dermal co-exposures for the pesticides Chlordane, Chlorpyrifos, and Dieldrin (as noted above), as well as for nickel. The presence of the blood-nickel biomarker near the bottom of the list of estimated lambda ratios ($\hat{\lambda}_{bio} / \hat{\lambda}_{air} = 5.73$) shows that biomarker levels for a metal will not always be less variable than the corresponding air levels in an exposed population. Regarding the large estimated lambda ratio for the serum albumin adduct of styrene oxide in the population of reinforced-plastics workers, background sources of the same adduct were observed in unexposed subjects, thus reducing the observed variability in adduct levels across subjects in the population (Yeowell-O'Connell et al., 1996).

Table 12.3 Estimated lambda ratios ($\hat{\lambda}_{bio} / \hat{\lambda}_{air}$) for air and biomarker measurements [compiled from Lin et al. (2005)].

Agent	Biomarker	Medium	$\hat{\lambda}_{bio} / \hat{\lambda}_{air}$ (Median)	$\hat{\lambda}_{bio} / \hat{\lambda}_{air}$ (Range)	$\hat{\lambda}_{bio} / \hat{\lambda}_{air}$ (N)
Lead	Lead	Blood	0.031	0.017-0.046	2
Lead	Lead	Urine	0.039		1
Mercury	Mercury	Blood	0.049		1
Lead	δ-ALA	Urine	0.062		1
PERC	PERC	Breath	0.096	0.055-0.311	3
Et. oxide	N-2-HEV	Blood	0.107		1
Chromium	Chromium	Urine	0.112	0.111-0.114	2
MC	MC	Breath	0.169	0.060-0.277	2
p-DCB	p-DCB	Breath	0.379	0.296-0.439	3
Chloroform	Chloroform	Breath	0.425	0.017-0.833	2
TCE	TCE	Breath	0.432	0.209-0.654	2
Benzene	Benzene	Breath	0.487	0.080-0.740	4
o-Xylene	o-Xylene	Breath	0.721	0.092-1.35	2
Terpenes	Verbenol	Urine	0.818		1
Et. benzene	Et. benzene	Breath	0.905	0.303-1.51	2
Styrene	HPRT	Blood*	1.20		1
Styrene	Styrene	Breath	1.33	0.472-3.56	3
Heptachlor	Hept. epoxide	Blood	1.46		1
Styrene	Mandelic acid	Urine	1.71	1.33-2.09	2
Styrene	Styrene	Blood	1.80	0.448-3.16	2
Nickel	Nickel	Urine	2.05	0.373-3.76	2
SO	SCE	Blood*	2.94		1
Nickel	Nickel	Blood	5.73		1
Dieldrin	Dieldrin	Blood	7.22		1
Styrene	Styrene glycol	Blood	7.51		1
Chlordane	Oxychlordane	Blood	17.3		1
Chlorpyrifos	TCPY	Urine	23.0		1
SO	Alb. adduct	Blood	26.5		1
SO	DNA adduct	Blood*	46.0		1

Legend: N, number of estimated values of lambda ratios; δ-ALA, δ-aminolevulinicacid; p-DCB, p-dichlorobenzene; N-2-HEV, N-2-hydroxyethylvaline (hemoglobin adduct); Et. benzene, ethyl benzene; Et. oxide, ethylene oxide; Hept. epoxide, Heptachlor epoxide; HPRT, hypoxanthine guanine phosphoribosyl transferase (mutation frequency); MC, methyl chloroform; PERC, perchloroethylene; SCEs, sister-chromatid exchanges (in lymphocytes); SO, styrene-7,8-oxide; TCE, trichloroethylene; TCPY, 3,5,6-trichloro-2-pyridinol.

* Lymphocytes.

12.4 Choosing between air measurements and biomarkers

With the above results in mind, we now suggest a decision-making scheme, shown in Figure 12.4, to guide the selection between use of air measurements and biomarkers for an epidemiologic study. The first step is to determine whether exposure occurs primarily via inhalation. If inhalation is not known to be the primary route of exposure, then we believe that biomarkers should be used as surrogates for true exposure levels. If, on the other hand, inhalation is known to be the primary route of exposure, then preliminary data should be used to estimate the lambda ratio ($\hat{\lambda}_{bio} / \hat{\lambda}_{air}$) for any candidate biomarker. If the estimated lambda ratio is not substantially less than one, say $\hat{\lambda}_{bio} / \hat{\lambda}_{air} > 0.75$, then air measurements would generally be chosen as exposure surrogates, due to their ease of collection and acceptance by subjects. But, if the estimated lambda ratio is substantially less than one, i.e., $\hat{\lambda}_{bio} / \hat{\lambda}_{air} \leq 0.75$, then biomarkers appear to be less biasing surrogates than air measurements and may be preferred for investigating health effects. However, because biomarkers tend to be more difficult and expensive to obtain and analyze than air measurements, sample sizes and potential biasing effects should be investigated for both air samples and biomarkers, using information about costs and about the desired power to detect a meaningful exposure-response relationship. The sample size calculations provided in Chapter 10 (e.g., Section 10.3) should be useful for determining the numbers of measurements needed to ensure no more than a stated level of bias. If adequate sample sizes can be obtained with a biomarker, then biological monitoring would be preferred. Otherwise, air measurements might be able to achieve the desired outcome by relying upon larger numbers of repeated measurements per person to offset their greater biasing potential.

Using the compilation of estimated lambda ratios in Table 12.3 as a guide, 13 of the 29 biomarkers had median values of $\hat{\lambda}_{bio} / \hat{\lambda}_{air} \leq 0.75$, the cutoff value we recommend for the decision-making scheme in Figure 12.4. Interestingly, about half of these are short-term biomarkers, representing exhaled breath levels of several volatile organic compounds (namely, perchloroethylene, methyl chloroform, *p*-dichlorobenzene, chloroform, trichloroethylene, benzene, and *o*-xylene). This suggests that breath sampling offers a simple and inexpensive alternative to air sampling for investigations involving volatile organic compounds. The other examples with values of $\hat{\lambda}_{bio} / \hat{\lambda}_{air} \leq 0.75$ were all very slowly eliminated biomarkers, including several metals (or their products) in blood or urine and the hemoglobin adduct of ethylene oxide.

The 16 biomarkers with values of $\hat{\lambda}_{bio} / \hat{\lambda}_{air} > 0.75$ in Table 12.3 would appear to offer relatively little advantage over air measurements as surrogates for true exposure levels unless other compelling arguments can be made. For example, we have alluded to the possibility that exposures involving pesticides and nickel may not arise primarily from air, in which case our decision-making scheme indicates that biomarkers would be preferred as surrogates for true exposure levels. Assuming this to be the case, then 18 of the 29 biomarkers in Table 12.3 would be preferred to air measurements as exposure surrogates in epidemiologic studies. To the extent that the contaminants listed in Table 12.3 are representative of those for which health effects would be investigated, then our results indicate that more than half of occupational or environmental epidemiologic studies should consider using biomarkers rather than air measurements as surrogates for true exposure levels.

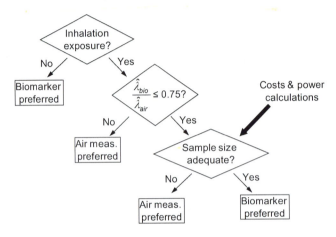

Fig. 12.4 Proposed decision-making scheme for selecting between the use of air measurements and a biomarker for an epidemiologic study. Note that the estimated lambda ratio, $\hat{\lambda}_{bio}/\hat{\lambda}_{air}$, is the estimated variance ratio for the biomarker divided by the estimated variance ratio for air measurements.

12.5 This chapter

In this chapter, we introduced biomarkers and discussed their connection to the exposure-disease model presented in Chapter 11 (Figure 11.1). We showed that biomarkers are remarkably diverse, both in terms of their positions in the exposure-disease continuum and in terms of their functions (biomarkers of exposure, effect, and susceptibility). We found it convenient to define three categories of biomarkers according to their residence times, namely, short-term biomarkers with residence times of two days or less, intermediate-term biomarkers with residence times between two days and two months, and long-term biomarkers with residence times greater than two months.

We then considered biomarkers as surrogates for exposure in epidemiologic studies and compared their characteristics to those of air measurements. We showed that biomarkers typically have smaller estimated variance ratios than air measurements in a given population, suggesting that biomarkers would be less biasing surrogates of true exposure levels in a study of exposure-related health effects. This was particularly true for intermediate-term and long-term biomarkers, which effectively damp exposure variability and thus tend to have smaller within-subject variance components than corresponding air measurements. Overall, our results indicated that biomarkers would be preferred to air measurements for over half of the contaminants investigated.

Bibliography

ACGIH. *Documentation of the Threshold Limit Values and Biological Exposure Indices, 7th Ed.* Cincinnati, Ohio: American Conference of Governmental Industrial Hygienists; 2001.

ACGIH. *2007 Threshold Limit Values and Biological Exposure Indices for Chemical Substances and Physical Agents.* Cincinnati, Ohio: ACGIH; 2007.

AIHA. "LOGAN Workplace Exposure Evaluation System User's Manual". Akron, Ohio: American Industrial Hygiene Association; 1990.

Aitchison J, Brown JAC. *The Lognormal Distribution.* London: Cambridge University Press; 1957.

Alavanja CR, Brown CR, Spirtas R, Gomez M. Risk assessment for carcinogens: A comparison of approaches of the ACGIH and the EPA. *Appl Ind Hyg*, 1990; 5: 510-517.

Albertini R, Clewell H, Himmelstein MW, Morinello E, Olin S, Preston J, Scarano L, Smith MT, Swenberg J, Tice R, Travis C. The use of non-tumor data in cancer risk assessment: reflections on butadiene, vinyl chloride, and benzene. *Regul Toxicol Pharmacol*, 2003; 37(1): 105-32.

Armstrong BG. Effect of measurement error on epidemiological studies of environmental and occupational exposures. *Occup Environ Med*, 1998; 55(10): 651-6.

Armstrong BK, White E, Saracci G. *Principles of Exposure Measurement in Epidemiology.* New York, NY: Oxford University Press; 1992.

Ashford JR. The design of a long-term sampling programme to measure the hazard associated with an industrial environment. *J Royal Stat Soc, Series A*, 1958; 3: 333-347.

Bakke B, Stewart P, Eduard W. Determinants of dust exposure in tunnel construction work. *Appl Occup Environ Hyg*, 2002; 17(11): 783-96.

Balsat A, de Graeve J, Mairiaux P. A structured strategy for assessing chemical risks, suitable for small and medium-sized enterprises. *Ann Occup Hyg*, 2003; 47(7): 549-56.

Bloomfield JJ, Greenburg L. Sand and metallic abrasive blasting as an industrial health hazard. *Journal of Industrial Hygiene*, 1933; 15: 184-204.

Box GEP, Jenkins GM. *Time Series Analysis: Forecasting and Control.* San Francisco: Holden-Day; 1976.

Breslin AJ, Ong H, Glauberman H, George AC, LeClare P. The accuracy of dust exposure estimates obtained from conventional air sampling. *Am Ind Hyg Assoc J*, 1967; 28: 56-61.

Brunekreef B, Noy D, Clausing P. Variability of exposure measurements in environmental epidemiology. *Am J Epidemiol*, 1987; 125(5): 892-8.

Buringh E, Lanting R. Exposure variability in the workplace: its implications for the assessment of compliance. *Am Ind Hyg Assoc J*, 1991; 52(1): 6-13.

Burstyn I, Teschke K. Studying the determinants of exposure: a review of methods. *Am Ind Hyg Assoc J*, 1999; 60(1): 57-72.

Burstyn I, Kromhout H. Are the members of a paving crew uniformly exposed to bitumen fume, organic vapor, and benzo(a)pyrene? *Risk Anal*, 2000; 20(5): 653-63.

Burstyn I, Kromhout H, Boffetta P. Literature review of levels and determinants of exposure to potential carcinogens and other agents in the road construction industry. *Am Ind Hyg Assoc J*, 2000a; 61(5): 715-26.

Burstyn I, Kromhout H, Kauppinen T, Heikkila P, Boffetta P. Statistical modelling of the determinants of historical exposure to bitumen and polycyclic aromatic hydrocarbons among paving workers. *Ann Occup Hyg*, 2000b; 44(1): 43-56.

Castleman BI, Ziem GE. Corporate influence on Threshold Limit Values. *Am J Ind Med*, 1988; 13: 531-559.

Cherrie JW. Are task-based exposure levels a valuable index of exposure for epidemiology? *Ann Occup Hyg*, 1996; 40(6): 715-22.

Cherrie JW. The beginning of the science underpinning occupational hygiene. *Ann Occup Hyg*, 2003; 47(3): 179-85.

Cherrie JW, Hughson GW. The validity of the EASE expert system for inhalation exposures. *Ann Occup Hyg*, 2005; 49(2): 125-34.

Clewell HJ, 3rd, Andersen ME, Barton HA. A consistent approach for the application of pharmacokinetic modeling in cancer and noncancer risk assessment. *Environ Health Perspect*, 2002; 110(1): 85-93.

Cochran WG. Errors of measurement in statistics. *Technometrics*, 1968; 10: 637-666.

Coenen W. The confidence limits for the mean values of dust concentration. *Staub Reinhalt Luft*, 1966; 26: 39-45.

Coenen W. Measurement assessment of the concentration of health-impairing, especially silicogenic dusts at work places of surface industries. *Staub Reinhalt Luft*, 1971; 31: 16-23.

Coenen W. Beschreibung des zeitlichen Verhaltens von Schadstoffkonzentrationen durch einen stetigen Markow-Process. *Staub Reinhalt Luft*, 1976; 36: 240-248.

Coenen W, Riediger G. Die Schatzung des zeitlichen Konzentrationsmittelwertes gefahrlicher Arbeitsstoffe in der Luft bei stichprobenartigen Messungen. *Staub Reinhalt Luft*, 1978; 38: 402-409.

Cohen B, McCammon CS. *Air Sampling Instruments, Ninth Edition.* Cincinnati, OH: ACGIH; 2001.

Cope R, Panacamo B, Rinehart WE, Ter Haar GL. Personal monitoring for tetraalkyllead in an alkyllead manufacturing plant. *Am Ind Hyg Assoc J*, 1979; 40: 372-379.

Corn M, Esmen NA. Workplace exposure zones for classification of employee exposures to physical and chemical agents. *Am Ind Hyg Assoc J*, 1979; 40(1): 47-57.

Corn M. Strategies of air sampling. *Scand J Work Environ Health*, 1985; 11: 173-180.

Corn M, Breysse P, Hall T, Chen G, T. R, Swift DL. A critique of MSHA procedures for determination of permissible respirable coal mine dust containing free silica. *Am Ind Hyg Assoc J*, 1985; 46: 4-8.

de Cock J, Heederik D, Kromhout H, Boleij JS, Hoek F, Wegh H, Ny ET. Determinants of exposure to Captan in fruit growing. *Am Ind Hyg Assoc J*, 1998; 59(3): 166-72.

Dempster AP, Laird NM, Rubin DB. Maximum likelihood from incomplete data via the EM algorithm. *J Royal Stat Soc, Series B*, 1977; 39: 1-38.

Dourson ML, Stara JF. Regulatory history and experimental support of uncertainty (safety) factors. *Reg Tox Pharmacol*, 1983; 3: 224-238.

Drinker P, Hatch T. *Industrial Dust.* New York: McGraw Hill; 1936.

Droz PO, Berode M, Wu MM. Evaluation of concomitant biological and air monitoring results. *Appl Occup Environ Hyg*, 1991; 6(6): 465-474.

Egeghy PP, Tornero-Velez R, Rappaport SM. Environmental and biological monitoring of benzene during self-service automobile refueling. *Environ Health Perspect*, 2000; 108(12): 1195-202.

Egeghy PP, Nylander-French L, Gwin KK, Hertz-Picciotto I, Rappaport SM. Self-collected breath sampling for monitoring low-level benzene exposures among automobile mechanics. *Ann Occup Hyg*, 2002; 46(5): 489-500.

Ehrenberg L, Moustacchi, E., Osterman Golkar, S. Dosimetry of genotoxic agents and dose-response relationships of their effects. *Mut Res*, 1983; 123: 121-182.

Esmen N, Hammad Y. Lognormality of environmental sampling data. *J Environ Sci Health*, *A12*, 1977: 29-41.

Esmen N. A distribution-free double sampling method for exposure assessment. *Appl Occup Environ Hyg*, 1992; 7: 613-621.

Evans JS, Hawkins NC. The distribution of student's t-statistic for small samples from lognormal exposure distributions. *Am Ind Hyg Assoc J*, 1988; 49: 512-515.

Francis M, Selvin S, Spear R, Rappaport S. The effect of autocorrelation on the estimation of workers' daily exposures. *Am Ind Hyg Assoc J*, 1989; 50(1): 37-43.

Galbas HG. Ein beschranktes sequentielles Testverfahren zur Beurteilung von Schadstoffkonztrationen am Arbeitsplatz. *Staub Reinhalt Luft*, 1979; 39: 463-467.

Gamble J, Spirtas R. Job classification and utilization of complete work histories in occupational epidemiology. *J Occup Med*, 1976; 18: 339-404.

George DK, Flynn MR, Harris RL. Autocorrelation of interday exposures at an automobile assembly plant. *Am Ind Hyg Assoc J*, 1995; 56: 1187-1194.

Greenburg L. Benzol poisoning as an industrial hazard. *US Pub Health Repts*, 1926; 41: 1516-1539.

Bibliography

Halton DH. A comparison of the concepts used to develop and apply occupational exposure limits for ionizing radiation and hazardous chemical substances. *Reg Tox Pharmacol*, 1988; 8: 343-355.

Hattis D. Pharmacokinetic principles for dose-rate extrapolation of carcinogenic risk from genetically active agents. *Risk Anal*, 1998; 18: 383-316.

Hawkins NC, Norwood SK, Rock JC. *A Strategy for Occupational Exposure Assessment.* Akron, Ohio: American Industrial Hygiene Association; 1991.

Heidermanns G, Kuhnen G, Riediger G. Messung and Beurteilung gesundheitsgefahrlicher Staube am Arbeitsplatz. *Staub Reinhalt Luft*, 1980; 40: 367-373.

Henschler D. Exposure limits: history, philosophy and future developments. *Ann Occup Hyg*, 1984; 28: 79-92.

Hewett P. Interpretation and use of occupational exposure limits for chronic disease agents. *Occup Med*, 1996; 11(3): 561-90.

Hewett P. Mean Testing: I. Advantages and disadvantages. *Appl Occup Environ Hyg*, 1997a; 12(5): 339-346.

Hewett P. Mean testing: II. Comparison of several alternative approaches. *Appl Occup Environ Hyg*, 1997b; 12(5): 347-355.

Hornung RW, Reed LD. Estimation of average concentration in the presence of nondetectable values. *Appl Occup Environ Hyg*, 1990; 5(1): 132-141.

Horvath E, Pongracz K, Rappaport S, Bodell WJ. ^{32}P-post-labeling detection of DNA adducts in mononuclear cells of workers occupationally exposed to styrene. *Carcinogenesis*, 1994; 15(7): 1309-1315.

Hughes JP. Mixed effects models with censored data with application to HIV RNA levels. *Biometrics*, 1999; 55: 625-629.

Janssen NA, Hoek G, Brunekreef B, Harssema H. Mass concentration and elemental composition of PM10 in classrooms. *Occup Environ Med*, 1999; 56(7): 482-7.

Johnson BA, Rappaport SM. On modeling metabolism-based biomarkers of exposure: a comparative analysis of nonlinear models with few repeated measurements. *Stat Med*, 2007; 26(9): 1901-19.

Jones RM, Nicas M. Margins of safety provided by COSHH Essentials and the ILO Chemical Control Toolkit. *Ann Occup Hyg*, 2006a; 50(2): 149-56.

Jones RM, Nicas M. Evaluation of COSHH Essentials for vapor degreasing and bag filling operations. *Ann Occup Hyg*, 2006b; 50(2): 137-47.

Juda J, Budzinski K. Fehler bei der Bestimmung der mittleren Staubkonzentration als Funktion der Anzahl der Einzelmessungen. *Staub Reinhalt Luft*, 1964; 24: 283-287.

Juda J, Budzinski K. Determining the tolerance range of the mean value of dust concentration. *Staub Reinhalt Luft*, 1967; 27: 12-16.

Koizumi A, Sekiguchi, T., Konno, M., Ikeda, M. Evaluation of the time weighted average of air contaminants with special references to concentration fluctuation and biological half time. *Am Ind Hyg Assoc J*, 1980; 41: 693-699.

Kromhout H, Oostendorp Y, Heederik D, Boleij JS. Agreement between qualitative exposure estimates and quantitative exposure measurements. *Am J Ind Med*, 1987; 12(5): 551-62.

Kromhout H, Symanski E, Rappaport SM. A comprehensive evaluation of within- and between-worker components of occupational exposure to chemical agents. *Ann Occup Hyg*, 1993; 37(3): 253-70.

Kromhout H, Swuste P, Boleij JS. Empirical modeling of chemical exposure in the rubber-manufacturing industry. *Ann Occup Hyg*, 1994; 38(1): 3-22.

Kromhout H, Heederik D. Occupational epidemiology in the rubber industry: implications of exposure variability. *Am J Ind Med*, 1995; 27(2): 171-85.

Kromhout H, Tielemans E, Preller E, Heederik D. Estimates of individual dose from current measurements of exposure. *Occup Hyg*, 1996; 3: 23-29.

Kromhout H, Vermeulen R. Temporal, personal and spatial variability in dermal exposure. *Ann Occup Hyg*, 2001; 45(4): 257-73.

Kromhout H, van Tongeren M. How important is personal exposure assessment in the epidemiology of air pollutants? *Occup Environ Med*, 2003; 60(2): 143-4.

Kumagai S, Matsunaga I, Kusaka Y. Autocorrelation of short-term and daily average exposure levels in workplaces. *Am Ind Hyg Assoc J*, 1993; 54(7): 341-50.

Kumagai S, Matsunaga I. Changes in the distribution of short-term exposure concentration with different averaging times. *Am Ind Hyg Assoc J*, 1995; 56(1): 24-31.
Kumagai S, Kusaka Y, Goto S. Cobalt exposure level and variability in the hard metal industry of Japan. *Am Ind Hyg Assoc J*, 1996; 57(4): 365-9.
Kumagai S, Kusaka Y, Goto S. Log-normality of distribution of occupational exposure concentrations to cobalt. *Ann Occup Hyg*, 1997; 41(3): 281-6.
Lagorio S, Iavarone I, Iacovella N, Proietto IR, Frselli S, Baldassarri LT, Carere A. Variation of benzene exposure among filling station attendants. *Occup Hyg*, 1998; 4: 15-30.
Land C. Hypothesis tests and interval estimates. In: Crow EL, Shimizu K, editors. *Lognormal Distributions*. New York: Marcel Dekker; 1988. p. 87-112.
Lange K. *Numerical Analysis for Statisticians*. New York: Springer-Verlag; 1999.
Lange N, Ryan LM. Assessing normality in random effects models. *Ann Stat*, 1989; 17: 624-642.
Langmead WA. Air sampling as part of an integrated programme of monitoring of the worker and his environment. *Inhaled Part Vapors*, 1970; 2: 983-995.
LaNier ME. Threshold Limit Values - Discussion and thirty-five year index with recommendations. *Ann Am Conf Ind Hyg*, 1984; 9(343-346).
LeClare PC, Breslin AJ, Ong LDY. Factors affecting the accuracy of average dust concentration measurements. *Am Ind Hyg Assoc J*, 1969; 30: 386-393.
Leidel NA, Busch K, Lynch J. "Occupational Exposure Sampling Strategy Manual", Report No. NIOSH 77-173. Washington, D.C.: U.S. Government Printing Office; 1977.
Liljelind IE, Stromback AE, Jarvholm B, Levin JO, Strangert BL, Sunesson A-LK. Self-assessment of exposure: a pilot study of assessment of exposure to benzene in tank truck drivers. *Appl Occup Environ Hyg*, 2000; 15(195-202).
Liljelind IE, Rappaport SM, Levin JO, Stromback AE, Sunesson AL, Jarvholm BG. Comparison of self-assessment and expert assessment of occupational exposure to chemicals. *Scand J Work Environ Health*, 2001; 27(5): 311-7.
Lin YS, Kupper LL, Rappaport SM. Air samples versus biomarkers for epidemiology. *Occup Environ Med*, 2005; 62(11): 750-60.
Lindstedt G, Gottberg I, Holmgren B, Jonsson T, Karlsson G. Individual mercury exposure of chloralkali workers and its relation to blood and urinary mercury levels. *Scand J Work Environ Health*, 1979; 5: 59-69.
Long WM. Airborne dust in coal mines: The sampling problem. *Br J Ind Med*, 1953; 10: 241-244.
Loomis DP, Peipins LA, Browning SR, Howard RL, Kromhout H, Savitz DA. Organization and classification of work history data in industry-wide studies: an application to the electric power industry. *Am J Ind Med*, 1994; 26(3): 413-425.
Lumens MEGL, Spee T. Determinants of exposure to respirable quartz dust in the construction industry. *Ann Occup Hyg*, 2001; 45(7): 585-595.
Lutz WK. Dose-response relationships in chemical carcinogenesis: superposition of different mechanisms of action, resulting in linear-nonlinear curves, practical thresholds, J-shapes. *Mutat Res*, 1998; 405(2): 117-24.
Lyles RH, Kupper LL. On strategies for comparing occupational exposure data to limits. *Am Ind Hyg Assoc J*, 1996; 57(1): 6-15.
Lyles RH, Kupper LL. A detailed evaluation of adjustment methods for multiplicative measurement error in linear regression with applications in occupational epidemiology. *Biometrics*, 1997; 53(3): 1008-25.
Lyles RH, Kupper LL, Rappaport SM. A lognormal distribution-based exposure assessment method for unbalanced data. *Ann Occup Hyg*, 1997a; 41(1): 63-76.
Lyles RH, Kupper LL, Rappaport SM. Assessing regulatory compliance of occupational exposures via the balanced one-way random effects ANOVA model. *J Ag Biol Environ Stat*, 1997b; 2(1): 64-86.
Lyles RH, Kupper LL. Measurement error models for environmental and occupational health applications. In: Rao CR, Sen PK, editors. *Chapter 17 in Handbook of Statistics, Vol 18: Bioenvironmental and Public Health Statistics*. Amsterdam, The Netherlands: Elsevier Science; 2000. p. 501-518.
Maxim LD, Allshouse JN, Venturin DE. The random-effects model applied to refractory ceramic fiber data. *Regul Toxicol Pharmacol*, 2000; 32(2): 190-9.
Mikkelsen AB, Schlunssen V, Sigsgaard T, Schaumburg I. Determinants of wood dust exposure in the Danish furniture industry. *Ann Occup Hyg*, 2002; 46(8): 673-85.

Bibliography

Mulhausen JR, Damiano J. *A Strategy for Assessing and Managing Occupational Exposures.* Washington, D.C.: AIHA Press; 1998.

Nicas M, Spear RC. A task-based statistical model of a worker's exposure distribution: Part II--Application to sampling strategy. *Am Ind Hyg Assoc J*, 1993; 54(5): 221-7.

Nieuwenhuijsen MJ, Lowson D, Venables KM, Newman-Taylor AJ. Correlation between different measures of exposure in a cohort of bakery workers and flour millers. *Ann Occup Hyg*, 1995; 39(3): 291-298.

NRC. *Biologic Markers in Pulmonary Toxicology.* Washington, D.C.: National Academy Press; 1990.

Oldham P, Roach SA. A sampling procedure for measuring industrial dust exposure. *Br J Ind Med*, 1952; 9: 112-119.

Oldham P. The nature of the variability of dust concentrations at the coal face. *Brit J Ind Med*, 1953; 10: 227-234.

Olsen E, Laursen B, Vinzents PS. Bias and random errors in historical data of exposure to organic solvents. *Am Ind Hyg Assoc J*, 1991; 52(5): 204-11.

Olsen E. Analysis of exposure using a logbook method. *Appl Occup Environ Hyg*, 1994; 9(10): 712-722.

Olsen E, Jensen B. On the concept of the "normal" day: quality control of occupational hygiene measurements. *Appl Occup Environ Hyg*, 1994; 9(4): 245-255.

Olsen E. Effect of sampling on measurement errors. *Analyst*, 1996; 121: 1155-1161.

Olson WH, Cumming RB. Chemical mutagens: dosimetry, Haber's rule and linear systems. *J Theor Biol*, 1981; 91: 383-395.

Occupational Safety and Health Act of 1970. Public Law 91-596. 91st Congress, S. 2193. Dec. 29, 1970.

OSHA. Occupational exposure to benzene. *Fed Reg*, 1987; 52: 34460-34579.

OSHA. American Iron and Steel Institute and Bethlehem Steel Corp. vs. Occupational Safety and Health Administration. *Federal Reporter 2nd Series (DC Cir)*, 1991; 939: 975-1010.

Paustenbach D, Langner R. Corporate occupational exposure limits: The current state of affairs. *Am Ind Hyg Assoc*, 1986; J. 47: 809-818.

Paustenbach DJ. The history and biological basis of occupational exposure limits for chemical agents. In: Harris RL, editor. *Patty's Industrial Hygiene and Toxicology, Fifth Edition.* New York, NY: John Wiley & Sons, Inc.; 2000. p. 1903-1999.

Peretz C, Goldberg P, Kahan E, Grady S, Goren A. The variability of exposure over time: a prospective longitudinal study. *Ann Occup Hyg*, 1997; 41(4): 485-500.

Peretz C, Goren A, Smid T, Kromhout H. Application of mixed-effects models for exposure assessment. *Ann Occup Hyg*, 2002; 46(1): 69-77.

Post W, Kromhout H, Heederik D, Noy D, Duijzentkunst RS. Semiquantitative estimates of exposure to methylene chloride and styrene: the influence of quantitative exposure data. *Appl Occup Environ Hyg*, 1991; 6(3): 197-204.

Preat B. Application of geostatistical methods for estimation of the dispersion variance of occupational exposures. *Am Ind Hyg Assoc J*, 1987; 48: 877-884.

Preller L, Heederik D, Kromhout H, Boleij JS, Tielen MJ. Determinants of dust and endotoxin exposure of pig farmers: development of a control strategy using empirical modeling. *Ann Occup Hyg*, 1995; 39(5): 545-57.

Rappaport SM. The rules of the game: an analysis of OSHA's enforcement strategy. *Am J Ind Med*, 1984; 6(4): 291-303.

Rappaport SM. Smoothing of exposure variability at the receptor: implications for health standards. *Ann Occup Hyg*, 1985; 29(2): 201-14.

Rappaport SM, Selvin S. A method for evaluating the mean exposure from a lognormal distribution. *Am Ind Hyg Assoc J*, 1987; 48(4): 374-9.

Rappaport SM, Selvin S, Roach S. A strategy for assessing exposure with reference to multiple limits. *Appl Ind Hyg*, 1988a; 3: 310-315.

Rappaport SM, Spear RC. Physiological damping of exposure variability during brief periods. *Ann Occup Hyg*, 1988; 32(1): 21-33.

Rappaport SM, Spear RC, Selvin S. The influence of exposure variability on dose-response relationships. *Ann Occup Hyg*, 1988b; 32 S1: 529-537.

Rappaport SM. Selection of the measures of exposure for epidemiology studies. *Appl Occup Environ Hyg*, 1991a; 6: 448-457.

Rappaport SM. Assessment of long-term exposures to toxic substances in air. *Ann Occup Hyg*, 1991b; 35(1): 61-121.

Rappaport SM. Biological considerations in assessing exposures to genotoxic and carcinogenic agents. *Int Arch Occup Environ Health*, 1993a; 65(1 Suppl): S29-35.

Rappaport SM. Threshold Limit Values, Permissible Exposure Limits, and feasibility: the bases for exposure limits in the United States. *Am J Ind Med*, 1993b; 23(5): 683-94.

Rappaport SM, Kromhout H, Symanski E. Variation of exposure between workers in homogeneous exposure groups. *Am Ind Hyg Assoc J*, 1993; 54(11): 654-62.

Rappaport SM, Lyles RH, Kupper LL. An exposure-assessment strategy accounting for within- and between-worker sources of variability. *Ann Occup Hyg*, 1995a; 39(4): 469-95.

Rappaport SM, Symanski E, Yager JW, Kupper LL. The relationship between environmental monitoring and biological markers in exposure assessment. *Environ Health Perspect*, 1995b; 103 Suppl 3: 49-53.

Rappaport SM, Yeowell-O'Connell K, Bodell W, Yager JW, Symanski E. An investigation of multiple biomarkers among workers exposed to styrene and styrene-7,8-oxide. *Cancer Res*, 1996; 56(23): 5410-6.

Rappaport SM, Weaver M, Taylor D, Kupper L, Susi P. Application of mixed models to assess exposures monitored by construction workers during hot processes. *Ann Occup Hyg*, 1999; 43(7): 457-69.

Rappaport SM, Goldberg M, Susi P, Herrick RF. Excessive exposure to silica in the U.S. construction industry. *Ann Occup Hyg*, 2003; 47(2): 111-22.

Rappaport SM, Kupper LL. Variability of environmental exposures to volatile organic compounds. *J Expo Anal Environ Epidemiol*, 2004; 14(1): 92-107.

Rappaport SM, Kupper LL, Lin YS. On the importance of exposure variability to the doses of volatile organic compounds. *Toxicol Sci*, 2005; 83(2): 224-36.

Riediger G. Die Anwendung von Maximalen Arbeitsplatzkonzentrationen (MAK) nach der TRgA 402. *Staub Reinhalt Luft*, 1986; 46: 182-186.

Roach SA. A method of relating the incidence of pneumoconiosis to airborne dust exposure. *Brit J Ind Med*, 1953; 10: 220-226.

Roach SA. A more rational basis for air-sampling programs. *Am Ind Hyg Assoc J*, 1966; 27: 1-12.

Roach SA, Baier EJ, Ayer HE, L. HR. Testing compliance with Threshold Limit Values for respirable dusts. *Am Ind Hyg Assoc J*, 1967; 28: 543-553.

Roach SA. A most rational basis for air sampling programmes. *Ann Occup Hyg*, 1977; 20(1): 65-84.

Roach SA, Rappaport SM. But they are not thresholds: a critical analysis of the documentation of Threshold Limit Values. *Am J Ind Med*, 1990; 17(6): 727-53.

Rock J. A comparison between OSHA-compliance criteria and action-level decision criteria. *Am Ind Hyg Assoc J*, 1982; 43: 297-313.

Rosner B, Willett WC, Spiegelman D. Correction of logistic regression relative risk estimates and confidence intervals for systematic within-person measurement error. *Stat Med*, 1989; 8(9): 1051-69; discussion 1071-3.

Rothman N, Stewart WF, Schulte PA. Incorporating biomarkers into cancer epidemiology: a matrix of biomarker and study design categories. *Cancer Epidemiol Biomarkers Prev*, 1995; 4(4): 301-311.

Searle SR, Casella G, McCulloch CE. *Variance Components*. New York: John Wiley; 1992.

Seixas NS, Sheppard L. Maximizing accuracy and precision using individual and grouped exposure assessments. *Scand J Work Environ Health*, 1996; 22(2): 94-101.

Selvin S, Rappaport SM, Spear RC, Schulman J, Francis M. A note on the assessment of exposure using one-sided tolerance limits. *Am Ind Hyg Assoc J*, 1987; 48: 89-93.

Sherwood R. Realization, development and first applications of the personal air sampler. *Appl Occup Environ Hyg*, 1997; 12: 229-234.

Sherwood RJ, Greenhalgh DMS. A personal air sampler. *Ann Occup Hyg*, 1960; 2: 127-132.

Sherwood RJ. On the interpretation of air sampling for radioactive particles. *Am Ind Hyg Assoc J*, 1966; 27: 98-109.

Sherwood RJ. The monitoring of benzene exposure by air sampling. *Am Ind Hyg Assoc J*, 1971; 32: 840-846.

Smith JS, Mendeloff JM. A quantitative analysis of factors affecting PELs and TLVs for carcinogens. *Risk Anal*, 1999; 19(6): 1223-34.

Bibliography

Smyth HF, Smyth HF, Jr. Spray painting hazards as determined by the Pennsylvania and the National Safety Council surveys. *J Ind Hyg*, 1928; 10(6): 163-214.

Spear RC, Selvin S, Francis M. The influence of averaging time on the distribution of exposures. *Am Ind Hyg Assoc J*, 1986; 47(6): 365-8.

Spear RC, Selvin S, Schulman J, Francis M. Benzene exposure in the petroleum refining industry. *Appl Ind Hyg*, 1987; 2: 155-163.

Spear RC, Selvin S. OSHA's permissible exposure limits: regulatory compliance versus health risk. *Risk Anal*, 1989; 9(4): 579-86.

Spengler J, Schwab M, Ryan PB, Colome S, Wilson AL, Billick I, Becker E. Personal exposure to nitrogen dioxide in the Los Angeles Basin. *J Air Waste Manage Assoc*, 1994; 44(1): 39-47.

Steenland K, Deddens JA. A practical guide to dose-response analyses and risk assessment in occupational epidemiology. *Epidemiology*, 2004; 15(1): 63-70.

Stewart P, Stenzel M. Data needs for occupational epidemiologic studies. *J Environ Monit*, 1999; 1(4): 75N-82N.

Susi P, Goldberg M, Barnes P, Stafford E. The use of a task-based exposure assessment model (T-BEAM) for assessment of metal fume exposures during welding and thermal cutting. *App Occup Environ Hyg*, 2000; 15: 26-38.

Swuste P, Hale A, Pantry S. Solbase: a databank of solutions for occupational hazards and risks. *Ann Occup Hyg*, 2003; 47(7): 541-7.

Symanski E, Rappaport SM. An investigation of the dependence of exposure variability on the interval between measurements. *Ann Occup Hyg*, 1994; 38(4): 361-72.

Symanski E, Kupper LL, Kromhout H, Rappaport SM. An investigation of systematic changes in occupational exposure. *Am Ind Hyg Assoc J*, 1996; 57(8): 724-35.

Symanski E, Kupper LL, Hertz-Picciotto I, Rappaport SM. Comprehensive evaluation of long-term trends in occupational exposure: Part 2. Predictive models for declining exposures. *Occup Environ Med*, 1998a; 55(5): 310-6.

Symanski E, Kupper LL, Rappaport SM. Comprehensive evaluation of long-term trends in occupational exposure: Part 1. Description of the database. *Occup Environ Med*, 1998b; 55(5): 300-9.

Symanski E, Chang CC, Chan W. Long-term trends in exposures to nickel aerosols. *Am Ind Hyg Assoc J*, 2000; 61(3): 324-33.

Symanski E, Chan W, Chang CC. Mixed-effects models for the evaluation of long-term trends in exposure levels with an example from the nickel industry. *Ann Occup Hyg*, 2001; 45(1): 71-81.

Symanski E, Savitz DA, Singer PC. Assessing spatial fluctuations, temporal variability, and measurement error in estimated levels of disinfection by-products in tap water: implications for exposure assessment. *Occup Environ Med*, 2004; 61(1): 65-72.

Tait K. The workplace exposure assessment expert system (WORKSPERT). *Am Ind Hyg Assoc J*, 1992; 53(2): 84-98.

Taylor DJ, Kupper LL, Rappaport SM, Lyles RH. A mixture model for occupational exposure mean testing with a limit of detection. *Biometrics*, 2001; 57(3): 681-688.

Taylor DJ, Kupper LL, Johnson BA, Kim S, Rappaport SM. Statistical methods for evaluating exposure-biomarker relationships. *J Agric Bio Environ Stat*, in press.

Tielemans E, Kupper LL, Kromhout H, Heederik D, Houba R. Individual-based and group-based occupational exposure assessment: some equations to evaluate different strategies. *Ann Occup Hyg*, 1998; 42(2): 115-9.

Tischer M, Bredendiek-Kamper S, Poppek U. Evaluation of the HSE COSHH Essentials exposure predictive model on the basis of BAuA field studies and existing substances exposure data. *Ann Occup Hyg*, 2003; 47(7): 557-569.

Tomlinson RC. A simple sequential procedure to test whether average conditions achieve a certain standard. *Applied Statistics*, 1957; 6: 198-207.

Tornero-Velez R, Symanski E, Kromhout H, Yu RC, Rappaport SM. Compliance versus risk in assessing occupational exposures. *Risk Anal*, 1997; 17(3): 279-92.

Tornqvist M, Fred C, Haglund J, Helleberg H, Paulsson B, Rydberg P. Protein adducts: quantitative and qualitative aspects of their formation, analysis and applications. *J Chromatogr B Analyt Technol Biomed Life Sci*, 2002; 778(1-2): 279-308.

Travis LL, Land ML. Estimating the mean data sets with nondetectable values. *Environ Sci Technol*, 1990; 24(7): 961-962.
Tuggle RM. Assessment of occupational exposure using one-sided tolerance limits. *Am Ind Hyg Assoc J*, 1982; 43: 338-346.
Ulfvarson U. Limitations to the use of employee exposure data on air contaminants in epidemiologic studies. *Int Arch Occup Environ Health*, 1983; 52: 285-300.
Ulfvarson U. Assessment of concentration peaks in setting exposure limits for air contaminants at workplaces, with special emphasis upon narcotic and irritative gases and vapors. *Scand J Work Environ Health*, 1987; 13: 389-398.
van Tongeren MJ, Kromhout H, Gardiner K. Trends in levels of inhalable dust exposure, exceedance and overexposure in the European carbon black manufacturing industry. *Ann Occup Hyg*, 2000; 44(4): 271-80.
van Tongeren MJ, Gardiner K. Determinants of inhalable dust exposure in the European carbon black manufacturing industry. *Appl Occup Environ Hyg*, 2001; 16(2): 237-45.
Vermeulen R, de Hartog J, Swuste P, Kromhout H. Trends in exposure to inhalable particulate and dermal contamination in the rubber manufacturing industry: effectiveness of control measures implemented over a nine-year period. *Ann Occup Hyg*, 2000; 44(5): 343-54.
Vermeulen R, Bos RP, Kromhout H. Mutagenic exposure in the rubber manufacturing industry: an industry wide survey. *Mutat Res*, 2001; 490(1): 27-34.
Waters MA, Selvin S, Rappaport SM. A measure of goodness-of-fit for the lognormal model applied to occupational exposures. *Am Ind Hyg Assoc J*, 1991; 52(11): 493-502.
Weaver MA, Kupper LL, Taylor D, Kromhout H, Susi P, Rappaport SM. Simultaneous assessment of occupational exposures from multiple worker groups. *Ann Occup Hyg*, 2001; 45(7): 525-542.
Wenker MAM, Kezic S, Monster AC, de Wolff FA. Metabolic capacity and interindividual variation in toxicokinetics of styrene in volunteers. *Human Exper Toxicol*, 2001; 20: 221-228.
Woitowitz HJ, Schacke G, Woitowitz R. Ranking estimation of the dust exposure and industrial medical epidemiology. *Staub Reinhalt Luft*, 1970; 30: 15-18.
Woskie SR, Shen P, Eisen EA, Finkel MH, Smith TJ, Smith R, Wegman DH. The real-time dust exposures of sodium borate workers: examination of exposure variability. *Am Ind Hyg Assoc J*, 1994; 55(3): 207-17.
Wright BM. The importance of the time factor in the measurement of dust exposure. *Brit J Ind Med*, 1953; 10: 235-240.
Yeowell-O'Connell K, Jin Z, Rappaport SM. Determination of albumin and hemoglobin adducts in workers exposed to styrene and styrene oxide. *Cancer Epidemiol Biomarkers Prev*, 1996; 5(3): 205-15.
Zielhuis RL, van der Kreek FW. The use of a safety factor in setting health based permissible levels for occupational exposure I. A proposal. *Int Arch Occup Environ Health*, 1979a; 42: 191-201.
Zielhuis RL, van der Kreek FW. Calculation of a safety factor in setting health based permissible levels for occupational exposure II. Comparison of extrapolated and published permissible levels. *Int Arch Occup Environ Health*, 1979b; 42: 203-215.
Zielhuis RL, Noordam PC, Roelfzema H, Wibowo AAE. Short-term occupational exposure limits: A simplified approach. *Int Arch Occup Environ Health*, 1988; 61: 207-211.

Index

accumulation constant 142
ACGIH 9, 10, 11, 12, 13, 86
action level (AL) 14, 86, 89
adduct
    albumin 145, 152
    defined 128
    DNA .. 128, 129, 142, 144, 145, 150, 151, 152
    hemoglobin 145, 151, 153
    promutagenic 128, 144, 146
    protein 143
adduction
    rate of 128, 142
air sampling 1
American Conference of Governmental Industrial Hygienists .... *See* ACGIH
analysis of variance (ANOVA) ... 39, 45, 46, 47, 49, 58, 59
AR(1) process 33, 34
area sampling 1, 2, 3
    compared to personal sampling 4, 5
    illustrated 3
area under curve (AUC) 132
attenuation bias 111, 114, 115, 118, 120, 121, 124, 125, 147
autocorrelated exposure series 33
autocorrelation 33, 34, 146
autocorrelation function 33
balanced data 21
benzene 2, 13, 14, 15, 35, 41, 57, 86, 117, 133, 135, 152
    air sampling of 2
    exceeding the STEL of 94
    OSHA standard 16
    PELs 15
    self assessment of exposure 25
    TLVs 12
between-person variability .... *See* variance components, between person
bioactivation 111, 127, 142
biological monitoring 143, 145, 146, 153
biomarker 153
biomarkers 138, 143, 144, 145, 146, 147, 148, 150, 154
    as surrogates for exposure . 147, 149, 154
    classified by residence times 149
    compared to air measurements 145
    defined 143
    for epidemiology 147

lambda ratios of 152
physiological damping of 151
stratified by lambda ratio 151
stratified by residence time 148
time scales of 144
variance ratios of 149
boiler makers 20, 69, 70, 77, 78, 107
box and whisker plots 69
breath measurements 25, 140, 141, 143, 151, 153
breathing-zone sampling 2, 6
    defined 2
    illustrated 4
burden 128, 129, 130, 133, 137
    defined 127
    initial 131, 136
    integrated 142
    of parent compound .. 128, 129, 131, 133, 134, 136, 140, 142
    of reactive electrophile 129
    residual 134, 137
    steady-state 133
    time series of 136, 137, 141
cancer 127, 128, 142
    as a health outcome 127
    linear model of 15
    risk of 13, 14
carcinogen 12, 13, 14, 127, 130
cell turnover 128, 142
censoring 35
compliance 8, 22, 81, 82, 84
    testing 81, 82, 87, 88, 89, 95
    vs. health risk 87
compound symmetry 57
concave downward 111, 112
concave upward 10, 112
construction workers .. 20, 24, 69, 70, 72, 79
control bands 19
controls
    group-level 40, 106
    individual-level 106, 107
    optimization of 98
    workplace .. 18, 27, 76, 77, 79, 80, 97, 98, 106, 108, 146
covariance
    of logged exposure levels 64
    of paired exposure measurements 43
    of variance components 47

covariates. 23, 50, 51, 62, 63, 69, 70, 71, 75, 78, 97
cumulative exposure...... 17, 83, 84, 99, 131, 135, 136, 139
damage
    molecular 128
damaged cells 128, 129, 142
determinants of exposure .. 6, 54, 63, 67, 69, 99
detoxification 112, 127, 142, 144
DNA damage 128
dose 129, 141, 142
    administered 130
    biologically effective 130
    defined 129
    exposure 129
    exposure-specific 132
    internal 44
    linear range of 83
    long-term 83, 127, 135, 142
    mean 137
    mercury 139
    of parent compound .. 131, 135, 136, 137
    rate-dependence of 10
    related to burden time series 141
    styrene 140, 141, 151
    target 145
effect modifiers 144
elimination half time 137, 139
elimination rate...... 132, 135, 137, 142, 144, 146
epidemiology 111, 114, 121, 145, 147
exceedance 82, 84, 85, 87, 92, 95
    assessed via tolerance limits 90
    compared to overexposure 89, 90
    defined 81
    estimates of 86, 87
    maximum value of 92, 93
exposure zones 19
exposure-disease model 127, 128, 129, 141, 142, 143, 154
exposure-response relationship... 11, 15, 21, 22, 109, 111, 112, 124, 125, 127, 147, 153
    nonlinear 10, 111
feasibility 14, 15, 17
first-order autoregressive process See AR(1) process
fold-range 48
    defined 48
    estimated 41, 48, 50, 54
    predicted 51
geometric mean 30, 35, 36, 113
geometric standard deviation 30
group mean exposure 16, 46, 62, 72, 75, 76, 77, 78, 84, 104, 118
    predicted 76
    related to probabilities 92, 95, 102

testing 103
    to define acceptable exposure 91
group-based study .. 112, 119, 120, 121, 122
health outcome 19, 40, 97, 116, 124, 127
    continuous 111, 112, 113
    dichotomous 112, 122
health outcome model 113, 117
heterogeneity of exposures.... 39, 52, 79, 99, 106, 108, 109
highly-perfused tissues 132
histogram 27, 28, 29, 31
homogeneous exposure groups 19
homogeneous variance 42, 116
individual-based study .. 112, 113, 114, 120, 121, 148
industrial hygiene See occupational hygiene
inorganic lead 17, 27, 28, 30, 36, 57, 86, 137, 148
intraclass correlation 43, 49, 52, 88, 90, 102
iron workers 20, 69, 70, 77, 78, 107
lag 33
    for AR(1) process 34
lambda ratios 150
likelihood ratio
    statistic 65
    test 65, 66, 67, 73, 116
limit of detection 35
linear kinetics 129, 135, 141, 142
logistic model 123
lognormal distribution 27, 29, 35, 36, 37, 43, 44, 46, 48, 85, 93, 94, 135
matrix notation 55
mean 42, 43
    biased estimation of 35
    cumulative exposure 142
    exposure level. 30, 35, 39, 40, 43, 44, 46, 53, 64, 83, 85, 134, 139
    exposure level, uniformity of 52
    minimum variance unbiased estimator of 35
    of a health response 123
    of logged exposure levels. 30, 33, 35, 40, 41, 42, 43, 44, 62, 63, 64, 115, 118
    of P-burden 134, 136, 137, 138
    of random effects 41, 63
    sample 45
    testing against an OEL 91
mean residence time 134, 151
mean squares 46
measurement error. 109, 111, 112, 120, 123, 124, 125, 147
    model of 122
mercury 139, 143
metabolite 83, 111, 127, 128, 129, 130, 138, 142, 143, 145
microenvironmental sampling 5
misclassification error 111

Index

mixed model .. 55, 57, 58, 59, 60, 62, 63, 67, 69, 72, 79, 80
monomorphic group 45
normality 27, 29, 37, 61, 73
  of logged exposures 36
  of random-person effects 43, 59
  of residuals 60, 76
observational group ... 20, 21, 23, 29, 30, 35, 39, 40, 41, 42, 43, 44, 45, 49, 50, 52, 54, 64, 69, 80, 81, 84, 85, 86
  defined 19
Occupational Exposure Limits (OELs) . 8, 9, 10, 15, 17, 18, 52, 80, 81, 84, 86, 91, 94, 95, 104, 146
occupational hygiene 1
occupational hygienists 1, 6, 8, 22, 81
Occupational Safety and Health Act .... *See* OSH Act
Occupational Safety and Health Administration .... *See* OSHA
one way random effects model 40, 41, 55, 56, 57, 59, 60
OSH Act 6, 8, 9, 14
OSHA 14, 16, 17, 27, 81, 82, 86, 99
overexposure .... *See* probability of overexposue
parametric statistical models 29
parent compound ... 127, 129, 130, 131, 133, 134, 142, 144
passive elimination 127, 132, 140, 142
passive monitor 4, 23
Permissible Exposure Limits (PELs) . 14, 15
  creating 15
  interpretation of 16
personal environments 19, 40, 52, 97, 98, 106, 108
personal measurements .. 6, 7, 27, 28, 33, 49, 53, 54, 69, 111, 143
personal monitors ... 22, 34, 39, *See* personal sampling
personal sampling 3, 4, 21, 39, 84
  compared to area sampling 4
physiological damping ... 137, 138, 139, 140, 148, 149, 151
pipe fitters 20, 69, 70, 77, 78, 107
probability density function (PDF) ... 43, 44, 45
probability of overexposure ... 84, 85, 86, 87, 88, 89, 95, 97, 109
  defined 83
  estimates of 86, 87
  maximum value of 92, 93
  testing of 98
Proc MIXED . 58, 59, 65, 66, 67, 72, 75, 103
procarcinogen 127
profile plots 57, 58, 72
quantitative exposure assessment
  origins of 1

random effects 55, 56, 98, 99
  error 62, 63, 115
  group 115
  person 40, 43, 59, 62, 72, 115
  standardized 73, 74, 76
random error .... *See* random effects,error
random sampling 23, 34
reactive electrophile 127, 128, 129, 142, 144, 145
regression coefficient
  for health outcome study .. 112, 120, 123, 148
  representing fixed effect 63, 76
regression models 75
  of health effects 123
  of variance components 50
reliability coefficient 114, 119, 120
repair 112, 128, 129, 142, 144, 146
repeat variable 57, 58
repeated measurements .. 30, 39, 49, 72, 115, 153
residual risk 15
restricted maximum likelihood (REML) . 58, 59, 64, 86, 100
risk 14, 15, 16, 17, 22, 40, 44, 80, 81, 82, 83, 84, 87, 88, 89, 95, 97, 99, 108, 112, 127, 128, 142, 148
safety factors 18
sample sizes 21, 53, 92, 99, 101
  and attenuation bias 114, 119
  for compliance testing 82, 87, 95
  for epidemiologic studies 7
  for testing overexposure 101
  from industrial surveys 7
sampling strategies 6
saturable process 83, 112, 129
self assessment of exposure 23, 24, 25
Shapiro-Wilks W test 37, 73
Short-Term-Exposure Limit .... *See* STEL
significant risk 14, 15
  defined 14
similar exposure groups 19
skewness 27, 28, 37, 82
standardized estimators
  of random effects 59
stationarity 32, 134
statistical parameters 31, 36, 37, 47, 59
  ANOVA estimates of 58
  biased estimates of 33
  estimates of 64
  estimation of 34, 35, 36, 45, 46, 109
  fixed effects 55
  of exposure distributions 7, 29, 84, 86, 98, 105
  of full and reduced models 65, 73
  of lognormal distributions 30
  of stationary exposure series 32
steady state 130, 132, 133

defined 131
STEL 10, 14, 94
 related to group mean exposure 94
strategy for integrating exposure assessment
 and control 97, 98
styrene . 5, 24, 32, 41, 57, 86, 107, 117, 132, 133, 140, 141, 142, 143
styrene oxide 150, 151, 152
subject-specific mean exposure 106
TEAM study 4, 53, 54
Threshold Limit Values ....*See* TLVs
time series 32, 33, 34, 136, 137, 139, 141, 142
time-weighted average (TWA) 2, 27
TLV committee 11, 12
TLVs 9, 11, 12, 15, 17, 18, 94
 ages of 13
 health basis of 11
TLV-STEL 94
 defined 10
TLV-TWA 10, 11, 86, 94
 defined 9
toxicokinetic model 131
toxicokinetics 142
trends
 of exposure levels 13, 32, 33, 57, 147
 of long-term exposures 32
true mean exposure level 112, 113, 123
unbalanced data 21
uniformly exposed 40, 52
unweighted least-squares estimator 113, 118
uptake 132, 142, 144, 146
uptake rate 127, 130, 131, 132, 133, 136, 143
variance
 ANOVA estimate of 46
 biased estimation of 35
 estimated for Wald-type statistic 100
 homogeneous 29
 minimum variance unbiased estimator of 35
 of exposure levels. 30, 40, 43, 44, 46, 64, 134

of logged exposure levels. 30, 33, 35, 40, 41, 42, 43, 44, 64, 90, 102
of mean exposure levels 44
of *P*-burdens 134, 138
sample 36
weighted for estimating the mean 46
variance components. 43, 46, 48, 54, 56, 62, 63, 64, 66, 69, 72, 76, 78, 89, 121, 147, 150
affected by fixed effects 78
between group 115
between person 39, 43, 44, 46, 49, 97
cumulative distribution of 49
defined 41
estimating across groups 64
estimation of 47
for air measurements and biomarkers 148
for disinfection byproducts 53
for measurement error models 112
for VOCs 53
model predictions 50, 51
ranges of 49
related to exposure ranges 52
under full and reduced models 65, 66
within group 115
within person 42, 43, 44, 46, 151
variance ratio 114, 154
estimated for air measurements and biomarkers 148
variance-covariance matrix. 55, 56, 57, 114, 118, 147
volatile organic compounds (VOCs) 53, 54, 132, 143, 151, 153
Wald-type test 100, 101, 104, 105, 106
welder fitters 69, 70, 77, 78, 107
welding fumes 20, 21, 69, 70, 78, 79, 80, 86, 107
within-person variability .... *See* variance components, within person
working limits 17
worst-case sampling 8
 defined 22